Climate Change and Forest Management in the Western Hemisphere

Climate Change and Forest Management in the Western Hemisphere has been co-published simultaneously as *Journal of Sustainable Forestry*, Volume 12, Numbers 1/2 2001.

The *Journal of Sustainable Forestry* Monographic "Separates"

Below is a list of "separates," which in serials librarianship means a special issue simultaneously published as a special journal issue *and* as a "separate" hardbound monograph. (This is a format which we also call a "DocuSerial.")

"Separates" are published because specialized libraries or professionals may wish to purchase a specific thematic issue by itself in a format which can be separately cataloged and shelved, as opposed to purchasing the journal on an on-going basis. Faculty members may also more easily consider a "separate" for classroom adoption.

"Separates" are carefully classified separately with the major book jobbers so that the journal tie-in can be noted on new book order slips to avoid duplicate purchasing.

You may wish to visit Haworth's website at . . .

http://www.HaworthPress.com

. . . to search our online catalog for complete tables of contents of these separates and related publications.

You may also call 1-800-HAWORTH (outside US/Canada: 607-722-5857), or Fax 1-800-895-0582 (outside US/Canada: 607-771-0012), or e-mail at:

getinfo@haworthpressinc.com

Climate Change and Forest Management in the Western Hemisphere, edited by Mohammed H. I. Dore (Vol. 12, No. 1/2, 2001). *This valuable book examines integrated forest management in the Americas, covering important global issues including global climate change and the conservation of biodiversity. Here you will find case studies from representative forests in North, Central, and South America. The book also explores the role of the Brazilian rainforest in the global carbon cycle and implications for sustainable use of rainforests, as well as the carbon cycle and the valuation of forests for carbon sequestration.*

Mapping Wildfire Hazards and Risks, edited by R. Neil Sampson, R. Dwight Atkinson, and Joe W. Lewis (Vol. 11, No. 1/2, 2000). *Based on the October 1996 workshop at Pingree Park in Colorado,* **Mapping Wildfire Hazards and Risks** *is a compilation of the ideas of federal and state agencies, universities, and non-governmental organizations on how to rank and prioritize forested watershed areas that are in need of prescribed fire. This book explains the vital importance of fire for the health and sustainability of a watershed forest and how the past acceptance of fire suspension has consequently led to increased fuel loadings in these landscapes that may lead to more severe future wildfires. Complete with geographic maps, charts, diagrams, and a list of locations where there is the greatest risk of future wildfires,* **Mapping Wildfire Hazards and Risks** *will assist you in deciding how to set priorities for land treatment that might reduce the risk of land damage.*

Frontiers of Forest Biology: Proceedings of the 1998 Joint Meeting of the North American Forest Biology Workshop and the Western Forest Genetics Association, edited by Alan K. Mitchell, Pasi Puttonen, Michael Stoehr, and Barbara J. Hawkins (Vol. 10, No. 1/2 & 3/4, 2000). *Based on the 1998 Joint Meeting of the North American Forest Biology Workshop and the Western Forest Genetics Association, Frontiers of Forest Biology addresses changing priorities in forest resource management. You will explore how the emphasis of forest research has shifted from productivity-based goals to goals related to sustainable development of forest resources. This important book contains fascinating research studies, complete with tables and diagrams, on topics such as biodiversity research and the productivity of commercial species that seek criteria and indicators of ecological integrity.*

"There is clear emphasis on the genetics, genecology, and physiology of trees, particularly temperate trees. . . . These proceedings are also testimony to what does or should distinguish forest biology from other sciences: a focus on intra- and inter-specific interactions between forest organisms and their environment, over scales of both time and place." (Robert D. Guy, PhD, Associate Professor, Department of Forest Sciences, University of British Columbia, Vancouver, Canada)

Contested Issues of Ecosystem Management, edited by Piermaria Corona and Boris Zeide (Vol. 9, No. 1/2, 1999). *Provides park rangers, forestry students and personnel with a unique discussion of the premise, goals, and concepts of ecosystem management. You will discover the need for you to maintain and enhance the quality of the environment on a global scale while meeting the current and future needs of an increasing human population. This unique book includes ways to tackle the fundamental causes of environmental degradation so you will be able to respond to the problem and not merely the symptoms.*

Protecting Watershed Areas: Case of the Panama Canal, edited by Mark S. Ashton, Jennifer L. O'Hara, and Robert D. Hauff (Vol. 8, No. 3/4, 1999). *"This book makes a valuable contribution to the literature on conservation and development in the neo-tropics. . . . These writings provide a fresh yet realistic account of the Panama landscape." (Raymond P. Guries, Professor of Forestry, Department of Forestry, University of Wisconsin at Madison, Wisconsin)*

Sustainable Forests: Global Challenges and Local Solutions, edited by O. Thomas Bouman and David G. Brand (Vol. 4, No. 3/4 & Vol. 5, No. 1/2, 1997). *"Presents visions and hopes and the challenges and frustrations in utilization of our forests to meet the economical and social needs of communities, without irreversibly damaging the renewal capacities of the world's forests." (Dvoralai Wulfsohn, PhD, PEng, Associate Professor, Department of Agricultural and Bioresource Engineering, University of Saskatchewan)*

Assessing Forest Ecosystem Health in the Inland West, edited by R. Neil Sampson and David L. Adams (Vol. 2, No. 1/2/3/4, 1994). *"A compendium of research findings on a variety of forest issues. Useful for both scientists and policymakers since it represents the combined knowledge of both." (Abstracts of Public Administration, Development, and Environment)*

Climate Change
and Forest Management
in the Western Hemisphere

Mohammed H. I. Dore
Editor

Climate Change and Forest Management in the Western Hemisphere has been co-published simultaneously as *Journal of Sustainable Forestry*, Volume 12, Numbers 1/2 2001.

Food Products Press
An Imprint of
The Haworth Press, Inc.
New York • London • Oxford

Published by

Food Products Press®, 10 Alice Street, Binghamton, NY 13904-1580 USA

Food Products Press® is an imprint of The Haworth Press, Inc., 10 Alice Street, Binghamton, NY 13904-1580 USA.

Climate Change and Forest Management in the Western Hemisphere has been co-published simultaneously as *Journal of Sustainable Forestry,* Volume 12, Numbers 1/2 2001.

Cover design by Thomas J. Mayshock Jr.

Library of Congress Cataloging-in-Publication Data

Climate changes and forest management in the Western Hemisphere/Mohammed H. I. Dore, editor.
    p. cm.
    "Has been co-published simultaneously as Journal of sustainable forestry, volume 12, numbers 1/2 2001."
    Includes bibliographic references.
    ISBN 1-56022-077-5 (alk. paper)–ISBN 1-56022-078-3 (pbk.: alk. paper)
    1. Forest management–America. 2. Climate changes–America. I. Dore, M. H. I.
SD139.C65 2000
333.75'097–dc21
                                    00-063655

Indexing, Abstracting & Website/Internet Coverage

This section provides you with a list of major indexing & abstracting services. That is to say, each service began covering this periodical during the year noted in the right column. Most Websites which are listed below have indicated that they will either post, disseminate, compile, archive, cite or alert their own Website users with research-based content from this work. (This list is as current as the copyright date of this publication.)

Abstracting, Website/Indexing Coverage . . . . Year When Coverage Began

- *Abstract Bulletin* **1993**

- *Abstracts in Anthropology* **1993**

- *Abstracts on Rural Development in the Tropics (RURAL)* **1993**

- *AGRICOLA Database* **1993**

- *Biology Digest (in print & online)* **2000**

- *Biostatistica* **1993**

- *BUBL Information Service, an Internet-based Information*
  Service for the UK higher education community
  <URL: http://bubl.ac.uk/> **1995**

- *CNPIEC Reference Guide: Chinese National Directory*
  of Foreign Periodicals **1996**

- *Engineering Information (PAGE ONE)* **1998**

- *Environment Abstracts. Available in print–CD-ROM–on*
  Magnetic Tape. For more information check:
  www.cispubs.com **1993**

- *Environmental Periodicals Bibliography (EPB)* **1993**

- *FINDEX <www.publist.com>* **1999**

(continued)

Special Bibliographic Notes related to special journal issues (separates) and indexing/abstracting:

- indexing/abstracting services in this list will also cover material in any "separate" that is co-published simultaneously with Haworth's special thematic journal issue or DocuSerial. Indexing/abstracting usually covers material at the article/chapter level.
- monographic co-editions are intended for either non-subscribers or libraries which intend to purchase a second copy for their circulating collections.
- monographic co-editions are reported to all jobbers/wholesalers/approval plans. The source journal is listed as the "series" to assist the prevention of duplicate purchasing in the same manner utilized for books-in-series.
- to facilitate user/access services all indexing/abstracting services are encouraged to utilize the co-indexing entry note indicated at the bottom of the first page of each article/chapter/contribution.
- this is intended to assist a library user of any reference tool (whether print, electronic, online, or CD-ROM) to locate the monographic version if the library has purchased this version but not a subscription to the source journal.
- individual articles/chapters in any Haworth publication are also available through the Haworth Document Delivery Service (HDDS).

Climate Change and Forest Management in the Western Hemisphere

CONTENTS

ABOUT THE EDITOR

Mohammed H. I. Dore, DPhil, is Professor of Economics at Brock University, St. Catharines, Ontario, Canada. He has published several books including *Global Environmental Economics* (1999) and *The Macrodynamics of Business Cycles* (1993). His articles have appeared in *Theory and Decision*, the *Journal of Environmental Economics and Management*, *Environmental Ethics*, *Science and Society*, the *Journal of Comparative Economics* and many other scholarly publications.

He serves on the editorial boards of several journals, and was appointed by UNESCO to be an Editor of one of the volumes of the *UNESCO Encyclopedia of Life Support Systems*. His research interests include climate change in Canada, and global environmental change and its distributional consequences on the rich and the poor. He has collaborated with an Inter-American team of researchers from South, Central, and North America on the role of forests in mitigating global warming. He is a member of NEPAMA, an international consortium of research, based at the University of Brasilia.

An Introduction to Climate Change and Forest Management in the Western Hemisphere

Mohammed H. I. Dore

This collection of papers is concerned with a number of interesting aspects of forest policy in South, Central and North America. The first part covers integrated forest management that includes not only concern over biodiversity but also the impact of climate change on forest management through new international initiatives such as policies designed to fulfil emission reduction targets by adopting bilateral policies called "joint implementation." The second part is concerned with the role of the Brazilian rainforest in the global carbon cycle and the implications for the sustainable use of the rainforests. The final part again focuses on the carbon cycle and the valuation of forests for carbon sequestration and sustainable management of forests, but in the context of North America, where the threat of deforestation is just as real both in Canada and in Mexico.

The first paper, by Kulshreshtha and Dore, argues for an integrated framework for the development of a forest management policy that would be optimal both globally and locally. For insights into such integration, they draw on the experience of the Canadian Model Forest Program, with special reference to the Prince Albert Model Forest, located in Saskatchewan. The Canadian Model Forest Program recog-

Mohammed H. I. Dore is Professor of Economics, Brock University, St. Catherines, ON, Canada, L2T 2M5 (E-mail: dore@adam.econ.brockU.ca).

[Haworth co-indexing entry note]: "An Introduction to Climate Change and Forest Management in the Western Hemisphere." Dore, Mohammed H. I. Co-published simultaneously in *Journal of Sustainable Forestry* (Food Products Press, an imprint of The Haworth Press, Inc.) Vol. 12, No. 1/2, 2001, pp. 1-5; and: *Climate Change and Forest Management in the Western Hemisphere* (ed: Mohammed H. I. Dore) Food Products Press, an imprint of The Haworth Press, Inc., 2001, pp. 1-5. Single or multiple copies of this article are available for a fee from The Haworth Document Delivery Service [1-800-342-9678, 9:00 a.m. - 5:00 p.m. (EST). E-mail address: getinfo@haworthpressinc.com].

1

nized various stakeholders who may have goals that are consistent with each other, or there may be goals of stakeholders that are in open conflict. The authors first review the obstacles to the development of integrated forest management policies, and suggest a concrete approach that they call "multicriteria evaluation of alternative management options." Such integration is possible only if all feedback effects of industrial activity in forests are taken into account. For example, timber harvesting may generate profits for some stakeholders, but may have a negative impact on the ecosystem and on climate, which in turn will affect human well being. This suggests that commercial profit by itself is a poor indicator for forest management. The authors then go on to suggest methods for aggregating the stakeholders' interests through consensus building and conflict resolution.

Montagnini and her associates consider the need to integrate an additional requirement for long-term sustainability, namely the need to maintain forest biodiversity. They draw on their experience with case studies in the subtropical humid forest reserve in Misiones, Argentina, as well as results of experiments carried out by CATIE in Costa Rica. The key question they ask is: in what way can the adverse impacts of harvesting on the maintenance of biodiversity be minimized?

From their Argentinean case study, they are able to conclude that the "uniform spacing" method of harvesting minimizes the loss of biodiversity. Three years after harvesting, the forest cut by uniform spacing had the highest density of seedlings of total as well as commercial species, and it also exhibited a high diversity of understorey plants and other trees.

Their CATIE case study in Costa Rica also yielded interesting lessons for forest management in the tropics; their work shows that low-density logging with a selection of remnant trees requires detailed pre- and post-harvest forest inventories to ensure successful regeneration and growth of desired species. Their results demonstrate the need for an adaptive approach to forest management, depending on location.

The section on forest management is rounded off by an article by Segura and Lindegaard on Joint Implementation in Costa Rica. "Joint implementation" in this context means the adoption of climate change policies, where countries that emit carbon cooperate with countries that have a large stock of forest (acting as a carbon sink). Costa Rica is the first country to sell Carbon Tradable Offset Bonds on the world market, and countries that have high carbon emissions can partly meet their CO_2 emission reduction targets by buying these bonds.

Joint Implementation is encouraged under Article 4.2 of the UN FCCC (1992) and endorsed by the Kyoto Protocol (1997). Segura and Lindegaard discuss the joint implementation agreement between Costa Rica and Norway in which Norway paid $2 million to Costa Rica for the sequestration of 200,000 tons of carbon in Costa Rican forests. The sequestration service will be provided over a 25 year period through reforestation, fire prevention and forest conservation projects in Costa Rica, as well as fostering community development that will enhance a more sustainable use of forests at the community level. The authors examine the impact of joint implementation in one community in Junquillal de Santa Cruz as a case study, and assess how the money from joint implementation is being used to encourage ecotourism, biodiversity conservation and the development of forest conservation.

Part two contains two papers that deal with the Amazonian rainforests of Brazil. The first, by Fearnside, draws attention to the large amount of carbon dioxide emissions due to deforestation in the Brazilian Amazon. These emissions are much larger than the official estimates issued by Brazilian government agencies. He also shows that most of the deforestation is caused by large and medium ranchers. Fearnside makes a strong case that the Global Environmental Facility (GEF) is wrong in giving credit only for saving forest that would otherwise be cut, but giving no credit at all for maintaining a stock of carbon in the form of a forest. Fearnside distinguishes the preservation of carbon stocks in the form of fossil fuels not used up from the stock in the form of preserved forests. While there are about 5000 Gigatons (Gt) of fossil fuels, the rate at which they are currently being burnt is comparatively small, namely about 6 Gt per year. While the adopt of efficiency measures and tax policy could in principle slow down the burning of fossil fuels, there are no comparable market based instruments that can curb or even slow down tropical deforestation–*the tropical forests are in danger of being burnt up within a century*. The world needs to recognize that not giving credit for the preservation of forests threatens not only biodiversity but also the world's hydrological cycle. In addition GEF needs to recognize that avoiding one ton of carbon emission now is more valuable than avoiding one ton later. Fearnside therefore makes a strong plea for discounting future benefits and emphasizing current benefits that will flow from preserving tropical rainforests.

In the second paper on rainforests, James Kahn and his co-authors do some fresh thinking on the sustainable use for the Amazon rainforests. They argue that if human disturbances mimic natural disturbances as closely as possible, ecosystem recovery would be assured. They consider various ways of forest harvesting, such as selective harvesting, using the "strip method," or using the "strip shelter belt" method. They then consider which of these methods comes closest to mimicking natural disturbances. After that, the authors turn their attention to the design of economic instruments that would induce harvesting behavior that would be consistent with an orderly recovery of the forest ecosystem.

The final part has three papers that cover the same themes as above but deal with North American forests. Dore and Johnston consider a thought experiment in which they put a value on Canadian forests for the ecological service of carbon sequestration. They treat forests as "joint" capital that absorbs carbon emissions produced by the manufacturing sector that uses fossil fuels. Using a mathematical model, they argue that the value of forests (for carbon sequestration only) would rise with emissions and fall with an increase in the total area forested. They then use auto-regressive integrated moving average (ARIMA) models to generate a dollar value of a hectare of forest for *one* ecological service provided by the forests, namely carbon sequestration. The results suggest ways of putting a lower bound on the stumpage value of forests for harvesting.

In the second paper, Tony Ward considers the history of forest harvesting policy in British Columbia (Canada), with a view to assessing how current policy has changed. The major policy change that occurred between 1987 and 1997 is that logging firms are now required to replant, and the cost of forest regeneration is now borne by the harvesting companies. The second major change that has occurred in B.C. is the reduction in the size of the clear-cut blocks. However, regeneration is left to the harvesters who prefer monoculture plantations, so that much of the rich biodiversity of the forests may be lost.

In the final paper Torres-Rojo and Flores-Xolocotzi consider deforestation and land use change in Mexico. They consider the sensitivities in land use changes between forestland, agriculture and livestock breeding. They find that livestock production yields higher returns than agricultural production per unit of land area, especially in forest-lands, which happen to be marginal for agriculture but not for stockbreeding. The effect on prices is quite different, since a price increment in the agricultural sector has a stronger effect on reducing forest cover than the same proportional price increment in the stockbreeding sector. The policy implication from these results is that an incentive program in the agricultural sector is more likely to produce more deforestation than and incentive program in the stockbreeding sector. An important implication of their results is that the stockbreeding sector leads to a faster rate of deforestation than the agricultural sector. In addition, forestlands converted to stockbreeding remain highly productive but forestlands converted to agriculture remain marginal. This means that the stockbreeding sector is a serious disincentive for conservation, since it makes it more desirable to convert forestlands. But unfortunately for forestry, increments in the value of forest products (produced in the forest) do not have a strong effect on mitigating deforestation. That means that additional activities which might increase the per hectare value of forestry such as the use of non-timber forest products, hunting, and recreation, among others, have almost no effect on reducing deforestation. If these results are valid, then they have important policy lessons for forest conservation in Mexico.

The collection of the papers in this volume reflects forest policies in the South, Central and North America. The papers show how forest policies are changing across the Western Hemisphere, but the threat to tropical forests in Central and South America remains undiminished. The emerging concern over climate change due to rising carbon dioxide emissions has brought a new importance to preserving the stock of forests in the Hemisphere. This has required a re-examination of both harvesting policy and land use policy to make it consistent with sustainable development in a global perspective.

PART I:
INTEGRATED FOREST MANAGEMENT

Integrated Forest Management:
Obstacles to a Comprehensive Integration
of Economic and Environmental
Dimensions

Surendra Kulshreshtha
Mohammed H. I. Dore

Surendra Kulshreshtha is Professor of Agricultural Economics, University of Saskatchewan, 3D12 Agriculture Building, 51 Campus Drive, Saskatoon, SK, Canada, S7N 5A8.

Mohammed H. I. Dore is Professor of Economics, Brock University, St. Catherines, ON, Canada, L2T 2M5 (E-mail: dore@adam.econ.brockU.ca).

An earlier version of this paper was presented at the Brasilia conference on "Dimensoes Humanas da Mudanca Clim Tica e do Manenjo Sustent vel das Florests das Americas: Uma Conferencia Interamericana–The Human Dimensions of Global Climate Change and Sustainable Forest Management in the Americas," University of Brasilia, December 1-3, 1997.

[Haworth co-indexing entry note]: "Integrated Forest Management: Obstacles to a Comprehensive Integration of Economic and Environmental Dimensions." Kulshreshtha, Surendra, and Mohammed H. I. Dore. Co-published simultaneously in *Journal of Sustainable Forestry* (Food Products Press, an imprint of The Haworth Press, Inc.) Vol. 12, No. 1/2, 2001, pp. 7-36; and: *Climate Change and Forest Management in the Western Hemisphere* (ed: Mohammed H. I. Dore) Food Products Press, an imprint of The Haworth Press, Inc., 2001, pp. 7-36. Single or multiple copies of this article are available for a fee from The Haworth Document Delivery Service [1-800-342-9678, 9:00 a.m. - 5:00 p.m. (EST). E-mail address: getinfo@ haworthpressinc. com].

SUMMARY. In forest management, actual applications of an integrated approach in forest management are not very frequent. Integrated forest management generally involves taking into account the totality of interactions of various sub-systems–social, economic, and ecological–within the biosphere, together with integration of goals set for such management. Several obstacles are identified in this paper to a successful integration of various sub-systems. Philosophical differences with respect to the rights of various members of these systems, fragmented and uni-disciplinary training, lack of scientific knowledge regarding the impact of one sub-system on the other, as well as lack of information on users and their goals, are some of the major obstacles. Lack of consensus when preferences of various users of the forest are heterogeneous further adds to the difficulties. Stakeholder participation in the management along with the use of multiple criteria decision-making tools offers a route to integrated forest management. *[Article copies available for a fee from The Haworth Document Delivery Service: 1-800-342-9678. E-mail address: <getinfo@haworthpressinc.com> Website: <http://www.haworthpressinc.com>]*

KEYWORDS. Integrated forest management, heterogeneous preferences, multiple criteria decision-making, valuation criteria

NEED FOR INTEGRATION

Ever since the report of the World Commission on the Environment and Development (see WCED), there is a growing recognition of long-term interdependencies that exist between environmental quality and human activities. The recognition is partly based on the acceptance of phenomena such as global climate change, depletion of the ozone layer, and other similar adverse impacts on the environment. The effects of global climate change are predicted to become apparent in the latter half of the 21st century, although there is disagreement about the magnitude and timing of such changes. Land based physical resources, such as forests and agriculture, will also be affected by the process of global climate change. These climatic changes have been predicted to have a significant impact on the well-being of society at large.[1]

In the past, environmental damage has been caused by a lack of explicit consideration of the environmental values of natural resources in conjunction with economic and other human values. Recognition of the impacts of the economic development process on the environment and of the corresponding long-term impact of such environmental modification on the economic system clearly indicates the necessity that environmental considerations should be an integral part of the decision-making process. It is the contention of this paper that although there are some obstacles to overcome in the application of an integrated approach, with the proper development of a conceptual framework, such an integration is possible. In the context of forest management, adoption of such an integrated framework should lead to the development of forest management policies that are optimal both locally and globally.

OBJECTIVES OF THIS PAPER

We begin by reviewing the concept of integration in the context of forest management, and explore its use in designing forest management policies in the Americas. Success of such integration in the context of the Canadian Model Forest Program, and in particular, the Prince Albert Model Forest is also reviewed. Based in part on that review, we give reasons why the integration has not been successful. This is followed by a description of a proposed method of integrating the environmental and socio-economic dimensions.

THE CONCEPT OF INTEGRATION:
AN INTRODUCTION

Let us start with the concept of integration itself. "Integrated resource management" is a term commonly used for an approach to managing resources that takes into account as many diverse values as possible. A literal definition of the term "integration" means *treating two or more things at the same level*. Although a definition of integrated forest management is not easily available, one could follow such a definition developed for another natural resource. Jeffery (1969) has defined integrated resource management as "the application of management strategies to achieve the maximum output from the optimized use of a given natural resource of a specific area for the benefit of a referent-group and its successors." In more recent periods, one also encounters a phrase "sustainable forest management," which according to Noss (1993), is not synonymous with sustaining forests. However, sustaining forests does require the integrated management of forest resources.

In the context of forest management, this integration involves taking into account dimensions of ecology and of economics, which according to Sadler (1995) is a new direction in thinking about the environment and development. Although environment and economic interactions are a fact of life, their joint consideration in the policy decisions still remains unintegrated. The two are often considered by separate departments or agencies of the policy makers. Norgaard (1988) has also made a plea for the development of a model which encompasses a broad understanding of human and environment interactions. This cannot be accomplished except through a very careful attention to detail, which might require the ranking of the most significant and crucial features of both the ecology and the ecological constraints as well as the pertinent behavioral characteristics of human agents and their short-term aspirations.

It is now necessary to define the exact nature of this integration. In the past, economic analyses have pursued two types of issues: (1) availability of natural resources and its effect on economic growth, and (2) the effect of economic activities on the quality of the environment. These issues are commonly defined as the field of Resource and Environmental Economics. Both of these types of analyses tend to provide only a one-way flow. As shown in Figure 1, this approach acknowledges an inter-relationship of the human system and the environmental system. But apart from this inter-relationship, it is assumed that the human system dominates and rests on the environment. Thus we have interfacing subsystems (top half of Figure 1).

In contrast, under an *integrated* analysis of human and environmental dimensions of resource use, all systems are studied through two-way feedback loops. All interactions among them are captured, and consequences, as measured through pertinent indicators, are evaluated.[2] According to Tollan (1992), a distinction should be made between *the environment* (which is the complex of air, water, land and living organisms surrounding humans), and *the ecosystem*, in which humans have an active part, and are indeed major actors. Thus, an *eco-systemic approach* will include social, economic, and environmental systems. Following Tollan (1992), such an approach will be *called eco-systemic approach to resource management*.

We now consider how such an eco-systemic approach can be applied to forest management. This can be attempted in at least three ways:[3] one, detailed consideration is given to various attributes of

FIGURE 1. Interfacing Sub-Systems versus Integrated Systems

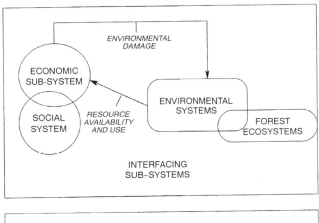

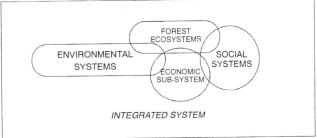

forests–trees, ground-cover, and non-timber products, both in terms of quality and quantity. Two, integration requires the recognition that forest systems interact with other systems, and such interactions must be explicitly taken into account in planning forest management. A third approach is to take a broader interpretation of forest ecosystems interacting with social and economic systems within the larger biosphere.

Although the first type of approach may be called "integrated" by some, it falls far short of the integrated resource management ideals based on the eco-systemic approach. Comprehensive integration would require an approach similar to the third mentioned above, in which forest resources are a part of a larger ecosystem, and a joint consideration of changes in them and the changes caused by them are recognized. *It is the totality of the interactions of the sub-systems within the biosphere that must form the database for planned resource management.*

Integrated resource management also requires the integration of a preliminary list of the goals of all "stakeholders." Some of these goals will be complementary and consistent. But some goals may be in open conflict. When there are conflicting goals, institutional mechanisms are required to try to achieve a common understanding. For example the Canadian Model Forest Program recognizes various stakeholders and attempts to provide a framework for consensus building. We do not suggest the building consensus is easy; indeed the process can be lengthy, as it requires furnishing data to all the stake holders and showing them feasible actions and possible consequences of current actions on future residents.

REVIEW OF PAST APPROACHES TO INTEGRATION

During the last two decades we have seen a rapid growth in the economic analysis of environmental and natural resource management issues. Such analyses are demanded not only by the bilateral and multi-national aid agencies, but also by national governments. Dixon (1995) indicates that the success in applying economic analysis to environmental problems has been mixed, partly because of the inherent limitations of economic theory, and partly because of the diversity of problems economics is being called upon to address.

According to Archibugi, Nijkamp, and Soeteman (1990), an important contribution to the economic–environmental integration is the principle behind the practice of *materials balance of resources* (extracted or collected, transformed, consumed and emitted), and the need to take into account the long-term social costs of such processes. Attempts have also been made (for example, Nordhaus and Tobin, 1972; and Fox, 1985) to build economic and social accounting systems which could incorporate measurement of economic welfare along with indicators of environmental quality. In addition, a number of evaluative methods such as extended benefit-cost analysis; environmental input-output models; total economic value of a natural or ecological resource; environmental impact assessment (EIA), among others have also been used to take into account some interactions between economic and environmental processes. Integration of economics and ecology has also been attempted from the viewpoint of land use, and pest management. All these approaches, to a certain degree, are oriented towards some limited measure of integration of economics and ecology.

In most of these approaches, (particularly the EIAs) one is interested in tracing the following interactions:[4]

1. Interdependencies among various economic agents within an economy;
2. Effects of economic activities on the environmental systems;
3. Effects of environmental systems (such as pollution damage, and loss of non-renewable resources) on the economic systems; and
4. Flows and interdependencies within various sub-systems such as diffusion of emissions, energy and material flows.

Although a comprehensive integrated approach must address all four of these interactions, one should also recognize the *time scale* in the generation of these interactions. The effects of economic activities on the environment may be more immediate,[5] while the impact of environment on the economic development and growth potential may be more in the long run. However, in practice, rules of thumb are developed and ecological services are valued by some ad hoc procedures. Finally both the economic and ecological services are added in monetary terms.

In order to generate the monetary valuation, three types of procedures are commonly used: One, use of the conventional markets; Two, use of implicit markets; and Three, use of artificial markets. The conventional market method uses market prices (with or without adjustments for market distortions) to changes in the economic activities. This method is applied for the production of market goods by the economy. The other two methods are used in the valuation of non-market goods. The basic idea behind the use of the implicit market method is to identify the links between the non-market goods and ordinary goods sold on the market. Methods exemplifying this method include: the Travel Cost Approach; Land and Property Value Approach; and Hedonic Price Models. The third method–the use of artificial market, bases inferences about value of non-market good from responses under some hypothetical market situation. These methods include the contingent valuation method (CVM) yielding estimates of willingness-to-pay (WTP) and willingness-to-accept (WTA). There are many critics of the use of the artificial market techniques.

A significant volume of literature exists evaluating the non-market valuation methods, and there appears to be no unanimity on the acceptance of a method of valuation among those available. However, in spite of the shortcomings of various methods, a report by the NOAA Panel (1993) has indicated:

- That passive use loss is a meaningful component of the total damage resulting from environmental accidents;
- That the contingent valuation (CV) studies are variable and some antagonists suggest that there can be no useful information in such studies; and
- That the CV studies convey useful information.

The major conclusion of the Panel was that CV studies can produce estimates reliable enough to be the starting point of a process of damage assessment, including lost passive values.

LIMITATIONS OF THE PAST APPROACHES TO INTEGRATION

Past approaches to integration have been based on two key cornerstones: (1) concept of value of a resource for efficient allocation; based on the concept of economic valuation, which has its roots in the neo-classical theory of value (relegated to Appendix 1); and (2) techniques that are used to integrate the human and the environmental dimensions. Both of these are examined in detail in the following section.

Concept of Value and Price for Efficient Allocation of Resources

Much of economic analysis is constructed around the single concept of economic efficiency. The idea here is very simple: the gains to society would be the greatest whenever there is a minimum of waste. However, economic efficiency can only be defined for a given distribution of wealth, resource rights, institutional arrangements, and the present generation. One of the necessary conditions for efficient resource allocation has been that all resources are bought and sold at their marginal social opportunity cost. In the context of natural resource how should this price be determined? In the neo-classical economic theory, this price would be equated to the marginal cost of supplying the resource to the users. However, in the context of efficient allocation from a sustainable use point of view, several other values would need to be included. Some of the noteworthy values are: (1) Marginal cost of any lost ecological functions; (2) Marginal cost of any pollution that the resource use may impose on other people; (3) Marginal cost of any lost future options; (4) Marginal cost of any lost existence and bequest values by the present generation; and (6) Marginal cost of resource use to the future generations.

Techniques of the Valuation of Environmental Damage

In addition to the concept of valuation, the validity of integration of human and environmental dimensions of resource use is also affected by the nature of technique used for the integration process. A review[6] of past studies suggest that many of the existing models, such as the input-output analysis, linear programming, benefit-cost analysis, and environmental impact assessment methods, for economic analyses are not without limitations. For example, input-output models assume a special production function, in which the cumulative impacts of smaller projects are disregarded. Furthermore, substitution between environmental degradation and international competitiveness between national industries is also not included in the decision making. These models can perhaps provide a total[7] environmental impact of various economic activities. However, the nature of the interaction is only one-way–from economic activities to the environmental quality, but without the two-way feedback referred to above.

Linear programming analysis of environmental tasks is not easy. Attempts have been made to incorporate the concerns of sustainable development of forests through such models (see McKillop and Sarkar, 1996). Problems such as the difficulty in incorporating joint costs, and economies of scale make it difficult to apply them to modeling of realistic situations.

Extended benefit-cost analysis is a very commonly used technique in the evaluation of economic activities with a significant impact on the environment. According to Bojo, Maler and Unemo (1992), the cost-benefit analysis is a normative analysis, aiding, but not dictating, environmental decisions. It is an input for policy makers. Furthermore, effects of the project on the environment are partially[8] included in the decision criteria, but the long-term effects of the environment upon economic development are excluded altogether. Similar to the EIA, the use of extended benefit-cost technique is for a project at a time, and thus, the cumulative effects are totally disregarded.

Environmental economics and resource economics have attempted to improve upon the shortcomings of the neoclassical economics, but these still suffer from serious limitations. According to Bossel (1996), the major impact of these disciplines is the inclusion of the monetary value of the external cost of all goods and services to society, the environment and future generations. However, empirical methods to estimate these externalities, as discussed above, are still replete with problems.

OBSTACLES TO SUCCESSFUL INTEGRATION

The integration of human and environmental dimensions of re-source use often faces many challenges. Some of these challenges can be labeled as obstacles in developing integrated resource management. These are discussed below.

Philosophical Differences

Efforts to put approaches to integrate economic and environmental dimensions into practice often run into entrenched interests groups and stakeholders. The range may be, as suggested by O'Riorden and Turner (1983) from technocentric to ecocentric (see Table 1). The former can be called "cowboy economics" mentality, where strong income growth orientation is required, which is possible only with a tendency towards unfettered resource exploitation. The other extreme is the ecocentric view of the world, which is based on deep ecology for preservation of all natural resources and a *de*-industrialization empha-sis in the development process. In fact, as O'Neill (1996) has stated, ecologists, with some notable exceptions, perpetuate the fantasy of a "natural world," where human society can be ignored.

Fragmentation at Policy Level as Well as at the Training Level

Much of the policy work is the responsibility of a line department with a governmental infrastructure. This means that the interrelation-ships among various impacts of the policy are either totally lost or minimized during the policy formulation stage. The very same bias

TABLE 1. Economic Development and Resource Management Paradigms

World View	Management Paradigm	Characteristics
Technocentric	Cowboy Economics	• Strong Growth Orientation • Tendency towards unfetted resource exploitation
	Managed Growth	• Recognition of environmental and social constraints to growth • Emphasis on efficient resource management
Bio-Economic	Sustainable Development	• Concerned with environmental–economic integration • Economic growth adjusted to bio-physical capacities and socio-cultural variations
Ecocentric	Human-Scale Ecodevelopment	• Biodiversity valued for its intrinsic worth • Emphasis on wilderness, protected areas, and species
	Deep Ecology	• Strong preservation orientation • Recognition of intrinsic values and rights for all animals • Emphasis on de-industrialization

Source: Based on O'Riordan and Turner (1983)

exists in the academic institutions with respect to the training and teaching of the interdisciplinary (or integrated) subjects. Although some interdisciplinary programs are emerging, most conventional programs at various educational institutions are still guided by the disciplinary bias.

Undiscovered Uses

Non-quantifiable impacts of development activities include changes in lifestyles, loss of historical or religious sites, and loss of genetic material that may have as yet undiscovered uses. In economic analyses, it is impossible to put a value on something that has not been proven to be a useful commodity.

Lack of Scientific Knowledge

Integration of human and environmental dimensions in resource use requires knowledge of impacts and changes in either economic activities as a result of environmental changes or vice-a-versa. Such knowledge is primarily based on the scientific data. There are some serious gaps in the science of resource management, both on the environmental aspects as well as in terms of economic activities. Such uncertainties created by the gaps in scientific knowledge slow down the adoption of integration of both dimensions.

Another major aspect of the gaps in our scientific knowledge is the long-run effect of changes in the environment and economic activities. For example, one of the long-run effects of environmental change is the changing global climate. Past efforts of the Intergovernmental Panel on Climate Change (IPCC) on developing knowledge on the long-run impacts on the economy have been sizeable. In spite of these, large gaps still remain in our understanding of these phenomena. There are many other issues, perhaps not as serious as the climate change, for which our scientific knowledge is relatively more poor.

Lack of Information on Use of Forests

Forest resources have multiple uses. For example, studies for the Prince Albert Model Forest region suggest that timber and non-timber uses are prevalent (see Table 2).

Results suggest that the nature of timber and non-timber uses of the forest are many. It also shows the fact that much of this information on the use and value of alternative sources of benefits from the forests is very poor. Relative to the timber uses, information on the non-timber uses is even poorer. Benefits also vary substantially from one use to the other.

Information on the commercial timber products is collected regularly, and is easily available. Furthermore, this information also enters explicitly in the private (commercial) and public decision making process. Information on the non-timber products, including the traditional uses of the forest, is not collected as a customary practice. Thus, this type of information may not enter explicitly into the decision making either at the private or public level.

Lack of Avenues for Consensus Building

Lack of information is related to the style of management of forest resources in a region. Possession of useful information is almost equivalent to holding some power over the others. In general situations, either the information is not available, or if available, it is not provided to all stakeholders. Information is key to informed decision-making, and a lack of this information may also lead to the absence of consensus building for the use of forest resources with multiple stakeholders.

TABLE 2. Magnitude of Use and Economic Value of Forests in Alternative Uses, Prince Albert Model Forest Region

Forest Use	Average Value per Permit Holder in Dollar	Number of Users
Timber Logging	32,742	271
Fuelwood Loggers– Non-Commercial	– 607	43
Fuelwood Loggers– Commercial	505	***
Pulp and Paper Producers (Workers)	– 104*	1 (1,163)
Sawmills Operators (Workers)	17**	1 (161)
Trappers	888	50
Outfitting	2,599	55
Specialty Forest Products Operators	700-70,500	9
Grazing Permit Holders	2,312	46
Wild Rice Producers	2,167	157/#
Commercial Fisherman	445	50

* per tonne of final product
** per thousand boardfeet of lumber
*** Included in non-commercial fuelwood loggers
\# Number of permit holders in Northern Saskatchewan
Adapted from Kulshreshtha (1995b)

In addition, lack of consensus may also come from the absence of a forum where various stakeholders can express their views, and listen to others about their interests in the management of the forest resource. For example, the recent creation of the Model Forest Network in Canada was undertaken in response to this need for integrated forest management.

Heterogeneity in Preferences

Various stakeholders do not have similar preferences. For example, Loewen and Kulshreshtha (1995b) based on a study of the Prince Albert Model Forest region suggested a significantly higher willingness-to-pay[9] for preservation of forest resources in the region, compared to the non-aboriginal people, as shown in Figure 2. Average willingness-to-pay for the aboriginal sample was around $82 per family, as against estimates of $43 to $75 for the non-aboriginal sample.

A similar conclusion can be drawn for the affiliation of stakeholders. For example, in the survey of Saskatchewan residents to estimate their value of wildland preservation, those with affiliations with Environmental Institutions had significantly higher WTP for most income groups, as shown in Figure 3.

FIGURE 2. Distribution of Average Willingness-to-Pay by Aboriginal and Non-Aboriginal Households by Income Levels and Zone of Residence

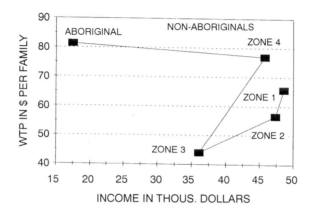

FIGURE 3. Distribution of Average Willingness-to-Pay by Income Class and Environmental Group Affiliation

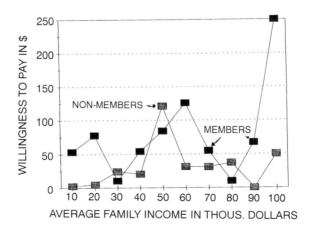

Preference for Selected Social Groups from Forest Management

Certain groups of people, such as the aboriginal people in Saskatchewan, depend upon forests for their traditional way of life. For example, a survey of the Montreal Lake Cree Nation located in the Prince Albert Model Forest Region suggested that forest is used for a variety of reasons, as shown in Table 3.

As shown in the Table, traditional uses of the forest are many and diverse. On average, 81% of the families at the Montreal Lake Cree Nation reserve had used the forest for traditional uses, which included hunting and trapping, fishing, and gathering of food products, medicines and craft materials. On average, families reporting such uses, spent an average of 54 days per annum on these activities.

A SUGGESTED APPROACH TO INTEGRATION

Comprehensive integration of economics and ecology to form a basis for environmental and economic policy making is badly needed. This has not been achieved so far, in spite of a number of strides that have been made in the past. This is not to suggest that no attempts have been made in the development of integrated economic–environment modeling. In fact a number of integrated models have been developed. A review of these models is provided in Braat and van Lierop (1987b). Some of the notable models include: land-marine

TABLE 3. Traditional Uses of Forest Resources in the Prince Albert Model Forest Region, 1993-94

Type of Use	% of All Families Reporting the Use	Average Number of Days per Family*
Hunting	82	32.5
Trapping	84	59.6
Fishing	91	71.7
Gathering Food	86	28.6
Medicines and Crafts	52	147.5
All Uses	81	53.7

* Average based on families reporting
Adapted from Kulshreshtha (1995b)

integrated development by Nishikawa et al. (1980); agricultural vegetation model by Moxey et al. (1995); and livestock production using
the indicators approach by Bouzaher et al. (1993).

The key question is: how is this integration to be achieved given the
uncertainty of scientific knowledge? An overview of the suggested
approach is shown in Figure 4. An adequate description of the system
is a key to make better informed decisions. Interactions among and
within each of the relevant systems have to be identified, since these
play an important role in trade-off analysis. Selection of the optimal
strategy would inevitably require multiple criteria. The proposed approach requires the following elements as its major elements:

1. Description of the Environmental and Economic Systems;
2. Interaction Between Economic and Environmental Systems;
3. Indicators for Human and Environmental Well-Being;
4. Estimation of Trade-Off Between Economic Growth and the Environmental Quality;
5. Consensus Building, Including Conflict Resolution, on the
   Goals and Objectives of Forest Management; and
6. Evaluation of Alternative Management Strategies.

Let us discuss these in the order presented.

Description of Economic and Environmental Systems

The nature of the description of the human (including economic and
other non-economic sub-systems) and the environmental system is
contingent upon the objectives of the forest management. A foremost
requirement for such a description is that a key theory needs to be built
using key concepts about the economy and the environment (Sproule-
Jones, 1995). Without theory, knowledge of the relationships between
the economy and the environment will remain merely descriptive
accounts.

The description of these systems must be based on the constructs of
systems analysis. According to Bossel (1996), many of the problems
society faces have interrelated ethical, cultural, interregional, ecological, economical, social, political, and technological aspects. Such
problems require the use of an integrated systems approach.

FIGURE 4. A Schematic Presentation of Integrated Forest Resource Management Model

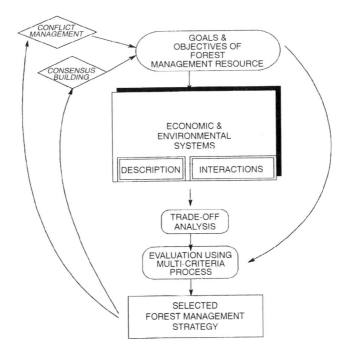

Interactions Between Economic and Environmental Systems

Interactions between environmental resources and human activities can be many and varied. Freeman (1979) has suggested two channels through which environmental changes affect human being and their activities. These are shown in Table 4.

The environment could affect economic and other human activities through living systems. Such effects are commonly felt in the commercial production, as well as in terms of human health. In addition to these, the environment could affect human activities through non-living systems, which include effects on climate as well as other aesthetic features.

Forests, like other natural resources, perform four types of functions: Regulation Function, Production Function, Carrier Function, and Information Function. These functional roles must be taken into account in developing the proper interaction between the ecosystems and human systems.

TABLE 4. Channels Through Which Environmental Changes Affect Human Beings

Channel	Effect on Human Being
Through Living Systems	1. Human Health
	2. Economic Productivity: such as Agriculture, Forestry, Fisheries, Recreation (Production), among others
	3. Other Ecosystem Impacts: such as Recreation (Amenities), Ecological Diversity, and Stability, among others
Through Non-Living Systems	4. Material Damage: such as Soiling, Production Costs
	5. Weather and Climate
	6. Other Effects: such as Odor, Visibility, and Visual Aesthetics

Adapted from Freeman (1979), p. 20

A conceptual framework for modeling the integrated resource management is shown in Figure 5. Here, both the environmental and human systems interact both within and between themselves. A comprehensive integration of human and environmental dimension in forest management must define connections between four element: Economy, Development, Environment, and Value System.

Indicators for Economic and Environmental Health of the Systems

De Groot (1995) has suggested four indicators for human welfare and national wealth, which are based both human-produced goods, and natural goods and services. In this scheme, a functional approach to natural resources is prescribed. Similarly, Simonis (1990) has suggested the use of indicators concerning material side of production, and environmentally relevant features of the production process. Use of the indicator approach is preferable for situations where there are multiple stakeholders, or every stakeholder has multiple objectives.

In the context of forest management a number of indicators can be identified. Based on the work by den Butte and Verbruggen (1994), three types of indicators are commonly used: Pressure Indicators, Impact Indicators, and Sustainability Indicators. The first type of indicators includes by-products of economic activities affecting the environment. These include, among others, emissions, discharges, extractions, and interventions. Impact Indicators reflect the impact of pressure indicators on the environment. These may include health impacts,

FIGURE 5. A Conceptual Framework for Integration of Human and Environmental Dimensions of Resource Use

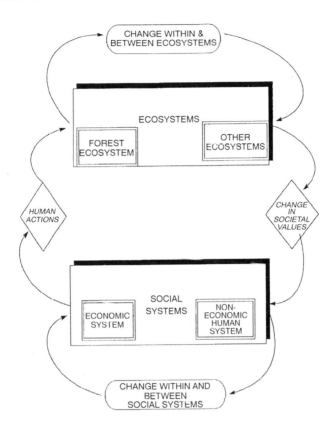

carrying capacity impacts, or limits to growth impacts. The Sustainability Indicators relate to the pressure indicators and impact indicators to some pre-selected goals of sustainability. The indicators have to be chosen in light of the objectives of the forest management and may be very region specific or group specific.

Trade-Off Analysis

Since most forest resource involve multiple stakeholders, and since most stakeholders, either individually or collectively have multiple objectives, use of a single criterion for evaluation of forest management strategies is not conceivable. Furthermore, on account of limited resources available at the disposal of the decision-makers, some goals may have to be sacrificed in favor of other goals. This requires the knowledge of trade-off functions for various goals. These functions would demonstrate to the decision-makers what must be sacrificed. At the macro-level, such trade-offs have been suggested to be shown by the natural resource accounting methods.

Consensus Building and Conflict Resolution

A major theoretical concern in environmental provision systems centers on rules for aggregating stakeholders' interests. Provisions may vary from highly consensual decision-making forums to highly coercive forums, dominated by particular interest groups (including government agencies). Negotiated decision-making, according to Dorcey (1995), occurs when individuals or groups make the trade-off themselves, and adopt an agreement based on their joint-decision making.

A partial measure of such trade-off between environment and economic development is presented through the natural resource accounting methods, which are just beginning to be used for environmental issues. In fact, Bojo, Maler and Unemo (1992) and den Butte and Verbruggen (1994) suggest natural resource accounting using a computable general equilibrium framework to study the interactions between economic development and environmental issues.

Multi-Criteria Evaluation
of Alternative Management Options

When stakeholders have multiple interests in a common resource, decision making by consensus is at best difficult and at worse, impossible. Several types of conflicts can be identified:

1. Conflicts between different groups within the society (economic development vs. environmental quality);
2. Conflicts between various regions of a country or the world (north-south conflicts, developed vs. developing country conflicts; or simply conflicts arising out of transboundary externalities);
3. Inter-temporal conflicts (backward conflicts arising out of preservation of heritage resources or forward conflict arising out of present use of resources); and
4. Inter-personal conflicts.

A partial explanation of the conflicts and their management in policy decision-making context is that objectives of the forest management tend to be multi-dimensional in nature. Therefore, the conventional economic optimizing tools may have some severe shortcomings to deal with these problems. Under these circumstances, multiple criteria (objectives or values) assessment methods deserve some consideration. According to Nijkamp (1990), the conflict nature of modern planning process and the society's desire not to force a single solution (as against a spectrum of solutions) have led to the popularity of multi-criteria evaluation method in the seventies onwards. According to Archibugi (1990), however, these methods are widely proposed at the abstract level, but have proven difficult to apply to the adequate political process of planning at the operational level.

Use of multi-criteria is based on a multi-attribute utility theory, which can be applied to discrete number of feasible solutions. The solutions are based on the trade-offs that exist for a given solution and the criteria set out by the decision-makers. These methods have been applied in agriculture and other management situations by Laxminarayan et al. (1991), and Romero and Rehman (1989). Compromise programming, one of the multiple criteria methods, for example, generates these trade-off functions between various policy goals.

CONCLUSIONS

Integrated forest resource management has been a widely used concept. In spite of its perceived need, a comprehensive integration of human and environmental dimensions has been partially successful. Many obstacles exist, including the economic valuation of the environment. As this paper has shown, the neoclassical concept of value is flawed and is unable to deal with issues related to the environment or sustainability and the rights of future generations. Perhaps, as Jakes (1992) has suggested, ecological economics is better suited to address these issues, since it proposes a much broader theory of value to include social, aesthetic, life support, intrinsic, and energy values, in addition to the traditional economic values. People's participation in the management, as suggested by Narayan (1995) is very important, as is the understanding of the trade-offs in selecting the final option. Multiple criteria decision-making tools offer some promise in tackling these issues.

NOTES

1. Among various impacts of climate change, food security and availability of water resources are considered to be the major ones. In addition, according to IPCC, incidence of various diseases, would also change. For details see, Munasinghe et al. (1996).

2. For more discussion on the integrated modeling, see Braat and van Lierop (1987a).

3. Mitchell (1990) discusses some of the approaches to integration in the context of water resources. But the application to integrated forest management has not been fully developed.

4. For a more detailed explanation of the integrated EIAs, see Whitney (1985).

5. According to van den Bergh (1993), this impact is an indirect result of two direct effects: the effect of development on the economy, and the effect of economic activities on the natural environment.

6. A review of economic models in the context of ecological systems is provided in Nijkamp (1987) and Pearse and Walters (1987).

7. The total impact on the environment is a sum total of direct, indirect and induced impacts. An input-output model is ideal for estimating these impacts.

8. The partiality of this inclusion stems from the fact that some of the environmental effects are non-quantifiable, and while others are non-monetizable.

9. These estimates are based on Loewen and Kulshreshtha (1995a) and Loewen and Kulshreshtha (1995b).

10. This is called the Pareto Optimality property of the General Competitive Equilibrium allocation.

11. It should be noted that the WTP is both individualistic and utilitarian.

12. When a departure occurs due to market power, it is called "time-tilting," compared to the benchmark path.

13. Other reviews of the theory can also be found in Edwards (1987), Gowdy and Olsen (1994), Sagoff (1988), and Willinger (1999).

REFERENCES

Archibugi, F. 1990. "Comprehensive Social Assessment: An Empirical Instrument for Environmental Policy Making," pp. 169-202, in F. Archibugi and P. Nijkamp (eds.), *Economy and Ecology: Towards Sustainable Development*. Boston: Kluwer Academic Publishers.

Archibugi, F., P. Nijkamp and F. J. Soeteman, 1990. "The Challenge of Sustainable Development," pp. 1-12, in F. Archibugi and P. Nijkamp (eds.), *Economy and Ecology: Towards Sustainable Development*. Boston: Kluwer Academic Publishers.

Bojo, J., K. Maler, and L. Unemo, 1992. *Environment and Development: An Economic Approach*. Boston: Kluwer Academic Publisher.

Bossel, H. 1996. "Ecosystems and Society: Implications for Sustainable Development," *World Futures*. 47:143-213.

Bouzaher, A., P. G. Laxminarayan, S. R. Johnson, T. Jones, and R. Jones, 1993. *The Economic and Environmental Indicators for Evaluating the National Pilot Project on Livestock and the Environment*. Staff Report 93-SR 64. Ames, Iowa: Center for Agricultural and Rural Development.

Braat, L. C. and W. F. J. van Lierop, 1987a. "Environment, Policy and Modeling," pp. 7-19, in L. C. Braat and W. F. G. van Lierop (eds.), *Economic-Ecological Modeling*. Amsterdam: North-Holland.

Braat, L. C. and W. F. G. van Lierop. 1987b. "Integrated Economic–Ecological Modeling," pp. 49-68, in L. C. Braat and W. F. G. van Lierop (eds.), *Economic-Ecological Modeling*. Amsterdam: North-Holland.

Chipman, J. and J. Moore, 1980. "Compensating variation, consumer's surplus and welfare," *American Economic Review.* 70: 933-949.

De Groot, 1995. "Environment Functions: An Analytical Framework for Integrating Environmental and Economic Assessment," pp. 55-80, in B. Sadler, E. Manning and J. Dendy (eds.). *Balancing the Scale: Integrating Environment and Economic Assessment*. Ottawa: Canadian Environmental Assessment Agency. April.

den Butte, F. A. G. and H. Verbruggen, 1994. "Measuring the Trade-off Between Economic Growth and Clean Environment," *Environmental and Resource Economics.* 4:187-194.

Dixon, J. A. 1995. "Experiences and Lessons from Environmental Applications of Economic Analytical Methods," pp. 37-54, in B. Sadler, E. Manning and J. Dendy (eds.). *Balancing the Scale: Integrating Environment and Economic Assessment*. Ottawa: Canadian Environmental Assessment Agency. April.

Dorcey, A. H. J., 1995. "Negotiation in the Integration of Environmental and Economic Assessment for Sustainable Development," pp. 101-118, in B. Sadler, E. Manning and J. Dendy (eds.). *Balancing the Scale: Integrating Environment and Economic Assessment*. Ottawa: Canadian Environmental Assessment Agency. April.

Dore, M. H. I., 1994. "Intergenerational Distribution of Exhaustible Resources: a Marshallian Approach." Paper presented at the Third Conference of the International Society for Ecological Economics, October 22-29, 1994, Costa Rica. Mimeo.

Dore, M. H. I., 1998. "Walrasian General Equilibrium and Nonlinear Dynamics," *Nonlinear Dynamics, Psychology, and Life Sciences*, Vol. 2, No. 1, pp. 59-72.

Dore, M. H. I. and G. C. Harcourt, 1986. "A Note on the Taxation of Resources under Oligopoly." *Economic Letters*, 21(1): 81-84.

Edwards, S., 1987. "In Defence of Environmental Economics," *Environmental Ethics.* 9: 75-85.

Freeman, M., 1979. *The Benefits of Environmental Improvement*. Baltimore: The Johns Hopkins Press.

Fox, K. A., 1985. *Social Systems Accounts: Linking Social and Economic Indicators through Tangible Behavioral Settings*. Dordrecht: Reidel.

Gowdy, J. M. and P. G. Olsen, 1994. "Further Problems with Neoclassical Environmental Economics," *Environmental Ethics.* 16: 161:171.

Hotelling, H. 1931. "The Economics of Exhaustible Resources," *Journal of Political Economy.* 39: 137-175.

Jakes, P. J., 1992. "What Does Ecological Economics Offer Natural Resource Management–A Proposed Research Program." Paper Presented at the Forestry and the Environment–Economic Perspective Conference, Jasper, March.

Jeffery, W. W. (Chairman), 1969. *Towards Integrated Resource Management*. Report of the Sub-Committee on Multiple Use, National Committee on Forest Land. Quebec, P. Q.

Kulshreshtha, S. N., 1995a. *Socio-Economic Baseline Survey of Montreal Lake Cree*

Nation. Report U of S: 95-06, Prince Albert: The Prince Albert Model Forest Association.

Kulshreshtha, S. N., 1995b. *Economics of Alternative Forest Uses: A Case Study of the Prince Albert Model Forest Region*. Report U of S: 95-09, Prince Albert: The Prince Albert Model Forest Association.

Laxminarayan, P. G., I. D. Atwood, S. R. Johnson and V. A. Spasio, 1991. "Compromise Solution for Economic-Environmental Decisions in Agriculture," *Journal of Environmental Management*. 33:51-64.

Loewen, K. G. and S. N. Kulshreshtha, 1995a. *Recreation and Wilderness: Participation and Economic Significance in Saskatchewan*. Report U of S: 95-02, Prince Albert: The Prince Albert Model Forest Association.

Loewen, K. G. and S. N. Kulshreshtha, 1995b. *Economic Aspects of Wilderness Valuation and Recreation Uses by Aboriginal Households: A Case Study of the Prince Albert*. Report U of S: 95-04, Prince Albert: The Prince Albert Model Forest Association.

Marshall, A., 1920. *Principles of Economics*. London: Macmillan Press.

McKillop, W. and A. Sarkar, 1996. Sustainable Development of Forest Resources in Industrialized Countries. *Agricultural Economics*. 14:175-184.

Mitchell, B., 1990. "Integrated Water Management," pp. 1-21, in B. Mitchell (Ed). *Integrated Water Management–International Experiences and Perspectives*. London: Belhaven Press.

Moxey, A. P., B. White, R. A. Sanderson and S. P. Rushton, 1995. "An Approach to Linking an Ecological Vegetation Model to an Agricultural Economic Model," *Journal of Agricultural Economics*. 46(3):381-397.

Munasinghe, M., P. Meier, M. Hoel, S. W. Hong, and A. Aaheim, 1996. Applicability of Techniques of Cost-Benefit Analysis to Climate Change," pp. 149-177, in Bruce, J. P., H. Lee, and E. F. Haites (Eds.), *Climate Change 1995–Economic and Social Dimensions of climate Change*. Cambridge: Cambridge University Press.

Narayan, D., 1995. *The Contribution of People's Participation: Evidence from 121 Rural Water Supply Projects*. Environmentally Sustainable Development Occasional Paper Series No. 1, Washington, DC: The World Bank.

Nijkamp, P. 1987. "Economic Modeling: Shortcomings and Perspectives," pp. 20-35, in L. C. Braat and W. F. G. van Lierop (eds.), *Economic-Ecological Modeling*. Amsterdam: North-Holland.

Nijkamp, P. 1990. "Multicriteria Analysis: A Decision Support System for Sustainable Environmental Management," in F. Archibugi and P. Nijkamp (eds.), *Economy and Ecology: Towards Sustainable Development*. Boston: Kluwer Academic Publishers.

Nishikawa, Y., S. Ikeda, N. Adachi, A. Udo, and H. Yukawa. 1980. "An Ecologic-Economic Model for Supporting Land-Marine Integrated Development–The Case of the East Seto Inland Sea," pp. 141-149, in Y. Haimes and J. Kindler (eds.). *Water and Related Land Resource Systems*. New York: Pergamon Press.

NOAA–National Oceanic and Atmospheric Administration Panel, 1993. "Natural Resource Damage Assessment under the Oil Pollution Act of 1990." *Federal Register*. 68(10): 4601-4614.

Nordhaus, W. D. and J. Tobin, 1972. "Is Growth Obsolete?" in National Bureau of Economic Research, *Economic Growth*. Columbia: Columbia University Press.

Norgaard, R. B., 1988. "Sustainable Development: A Co-Evolutionary Approach." *Futures.* 20(6):606-620.

Noss, R. F., 1993. "Sustainable Forestry or Sustainable Forests?" pp. 17-43, in G. H. Aplet, N. Johnson, J. T. Olson and V. A. Sample (eds.) *Defining Sustainable Forestry.* Washington, DC: Island Press.

O'Neill, R. V., 1996. "Perspectives on Economics and Ecology," *Ecological Applications.* 6(4):1031-1033.

O'Riorden, T. and R. K. Turner (eds.), 1983. *An Annotated Reader in Environmental Planning and Management.* Oxford, UK: Pergamon Press.

Pearse, P. and C. J. Walters, 1987. "Perspectives on the Application of Economic-Ecological Models," in L. C. Braat and W. F. G. van Lierop (eds.), *Economic-Ecological Modeling.* Amsterdam: North-Holland.

Rolston, III, Holmes,1985. "Valuing Wildlands," *Environmental Ethics.* 7: 23:48.

Rolston, III, Holmes,1994. "Value in Nature and the Nature of Value," pp. 13-30, in Robin Attfield and Andrew Belsey, (Eds.) *Philosophy and the Natural Environment.* Royal Institute of Philosophy Supplement: 36. Cambridge: Cambridge University Press.

Rolston, III, Holmes and J. Coufal. 1991. "A Forest Ethic and Multivalue Forest Management." *Journal of Forestry.* 89(4):35-40.

Romero, C. and T. Rehman, 1989. *Multiple Criteria Analysis for Agricultural Decisions.* New York: Elsevier.

Sadler, B. 1995. " Ecology, Economics and the Assessment of Sustainability: Themes and Approaches," pp. 1-36, in B. Sadler, E. Manning and J. Dendy (eds.). *Balancing the Scale: Integrating Environment and Economic Assessment.* Ottawa: Canadian Environmental Assessment Agency. April.

Sagoff, M. "Some Problems with Environmental Economics," *Environmental Ethics.* 10:55-74.

Simonis, U. E. 1990. "Ecological Modernization of Industrial Society–Three Strategic Elements," pp. 119-138, in F. Archibugi and P. Nijkamp (eds.), *Economy and Ecology: Towards Sustainable Development.* Boston: Kluwer Academic Publishers.

Sproule-Jones, M., 1995. "Institutional Design for the Economy and the Environment: the Identification and Representation of Stakeholders," pp. 85-100, in B. Sadler, E. Manning and J. Dendy (eds.). *Balancing the Scale: Integrating Environment and Economic Assessment.* Ottawa: Canadian Environmental Assessment Agency, April.

Tollan, A., 1992. "The Ecosystem Approach to Water Management," *World Meteorological Organization Bulletin.* 41(1):28-34.

van den Bergh, J. C. M., 1993. "A Framework for Modeling Economy–Environment–Development Relationships Based on Dynamic Carrying Capacity and Sustainable Development Feedback," *Environmental and Resource Economics.* 3:395-412.

Whitney, J. B. R., 1985. "Integrated Economic-Environmental Models in Environmental Impact Assessment," pp. 53-86, in V. M. Maclaren and J. B. Whitney (eds.). *New Directions in Environmental Impact Assessment in Canada.* Toronto: Methuen.

Willing R. 1976. "Consumer Surplus Without Apology," *American Economic Review.* 66: 589-597.

Willinger, M. 1999 "Non-Use Values and the Limits of Cost-Benefit Analysis," pp. 54-74 in Dore, M.H., Mount, T. H. (eds.) *Global Environmental Economics: Equity and the Limits to Markets* Cambridge, MA: Basil Blackwell.

WCED–The World Commission on Environment and Development, 1987. *Our Common Future*. Oxford: Oxford University Press.

APPENDIX 1

AN EXAMINATION OF ECONOMIC VALUATION CRITERION FOR ENVIRONMENTAL GOODS

Much of economic valuation of environmental goods, which are primarily non-market goods is done using various non-market valuation methods. These methods are developed such that they are expected to derive some proxy for market valuation. Let us examine the theoretical foundation of market valuation in detail.

CRITERION OF VALUATION IN NEOCLASSICAL ENVIRONMENTAL ECONOMICS

In the neoclassical economics, the concept of value depends on consumer demand, which expresses *the willingness to pay*. It arises quite unobtrusively from the Walrasian Model of General Equilibrium, on which all neoclassical theory is based. The entire economy is conceived as an auction, with participating and *willing* traders, who bring their *initial endowments* (of goods and labor) to the auction. In this (Walrasian) model, there is no production, since the initial endowments cannot be augmented.

In this auction, there is an assumed neutral auctioneer, who sets prices by attempting to balance supply (the initial endowments) and *conditional* demands for each and every commodity, i.e., conditional on a price. At this auction, no trade takes place until the auctioneer is satisfied that all supplies and demands for each and every commodity are in balance at some price. Since these series of intentions of trade are entirely voluntary (those who do not wish to buy or sell may withdraw at anytime during the auction, which is only a series of conditional expressions of trades), the final prices, at which all demands are in balance with supply, are called *competitive equilibrium prices*. All trade takes place only at these prices (and no other prices) in this Walrasian Model. These prices are based on the willingness-to-pay (WTP) of all traders. Notice that the auction reflects the wishes and valuations of only those people who are present at the auction, and those who have something to sell. Naturally, the future generations are not represented, except in so far as those present at the auction care for the rights and interests of the unborn.

In principle, all prices, including the prices of exhaustible resources, such as minerals in the ground, are determined by this auction, expressing the willingness to buy and to sell, or willingness to pay (WTP). *This is the criterion of valuation.* Thus, what would have occurred in a competitive auction (of given initial resources) is called the socially optimal allocation, because that final arrangement, discovered by the auctioneer by trial and error, cannot be improved upon.[10] Furthermore, this competitive allocation then becomes a social benchmark, for judging the desirability of all other allocations. Much of conventional environmental economics is concerned with valuing resources in such a way that will mimic, as closely as possible, the competitive equilibrium. Similarly, contingent valuation is an attempt to "discover" people's valuation, in such a competitive situation. Furthermore, all estimation procedures for determining the "value" or price of the so-called non-marketed goods (be it the value of the ozone layer or a wilderness area) rests on what is perceived to the willingness-to-pay for such amenities in the ideal competitive equilibrium conditions.

LIMITATIONS OF THE CONCEPT OF VALUATION

In the rest of this section, it is argued that the fundamental flaw in neoclassical environmental economics has to do with the criterion of valuation. All other neoclassical assumptions are of secondary importance.

1. Willingness to Pay (WTP) and Willingness to Accept (WTA)

In neoclassical economics, when there is a market for some good, the individualist and utilitarian assumptions of the theory dictate that the appropriate criterion of value is determined by "what the market will bear." In the economics textbooks, this is called *consumer willingness to pay* (or WTP for short). This *neoclassical* concept of value is diametrically opposed to the *classical* concept of value. The classical concept of value (in Smith, Ricardo, and Marx) is based on "real sacrifices," or more technically speaking, the social opportunity cost of producing a good. The neoclassical economists argue that cost is often a "sunk cost" and is not an adequate measure of the current value of a good. Only the current WTP is the appropriate guide to consumer preferences, and only WTP can guide production.[11]

The WTP, however, may be determined when a market for a good exists. But environmental economists are typically faced with an attempt to determine "market value" when no market exists for a good. For example, there is no market for the services of a forest. The question is how do we find a consumer's valuation of the forest. Neoclassical economists would now like to get to a close substitute: they would ask consumers, in survey questionnaires, *for the same good or amenity*, what consumers would be willing to pay for it. In the example of a forest, the consumers would be asked what they would be willing to pay to enjoy the beauty of the forest. A forest may contain a natural park which consumers would be willing to visit. They would be asked what would they be willing to pay per visit. This would be the WTP measure of the value of the forest. Its theoretical foundation lies in the Hicksian compensated demand curve. For a price fall, the Hicksian concept of value is the area under the Hicksain compensated demand curve, where the area is bounded by the two prices. This area is also called the Compensating Variation (CV).

But as there are two possible Hicksian compensated demand curves–one based on holding the utility constant at the original level, and the other at the new level after the price change, there is also an alternative to CV, which is called the Equivalent Variation (EV). In the example of the forest used above, it would involve asking the consumer what compensation he or she would desire for not being able to "visit" or view the beauty of the forest. This idea is what they would be *willing to accept* (or WTA for short) as compensation for the *loss* of the forest.

So far we have identified two measures of value. But of course there is a third, which is the Marshallian Consumer Surplus (CS), which is the area under the Marshallian demand curve, bounded by the two prices.

In general, it can be shown that for a price decrease CV < CS < EV, and for a price increase EV < CS < CV. An invariant measure of value would require a *unique measure of value*. That is, when CV = CS = EV. Unfortunately in neoclassical theory, a unique measure of value exists if and only if the underlying preferences are homothetic (see Chipman and Moore, 1987), which violates Engel's Law, perhaps the only true "law" in economics. All environmental economists who do empirical work on valuing environmental amenities ignore the problem of an invariant measure of value. Is there some workable empirical approach to valuation? That is, is there some way around this impasse? Willing (1976) has suggestion that CV and EV may be treated as the upper and lower bounds of CS. But that is clearly fudging the issue.

Thus the applied neoclassical economist has a choice: to obtain a WTP measure of value or to get a WTA as a measure. If they both turned out to be the same, then a unique measure of value would have been obtained. Unfortunately, most empirical studies show that the empirical measures of value of WTP and WTA do not agree. It turns out that in almost all actual empirical estimates, the WTA is higher than the WTP. The studies show that neoclassical theory is itself unable to reconcile the two different measures of value. Thus the neoclassical criterion of valuation is flawed: it cannot come up with a unique measure of value (see Dore, 1998). Given this difficulty, the neoclassical theorists ought to turn to the classical economists for the determination of value.

2. Cost versus Demand Determined Value

It was stated above that neoclassical economists argue that the classical doctrine of value being determined by *cost* is wrong, that value depends on what the market will fetch for something. Now, it is not well-known that Marshall reconciled this problem, essentially by saying that if you are talking about the value of fresh fish in the market, then it is the short-term value of the fish that is relevant (i.e., the WTP). Within a day the fish will go off and have no market value. Thus, in this short time horizon, value is determined by market price, or WTP, since the fish must be sold at whatever price the buyers are willing to pay. But the longer the time horizon, the influence of cost (actually social cost) must dominate. To quote Marshall (1920, p. 349):

> . . . (W)e may conclude that, as a general rule, the shorter the period which we are considering, the greater must be our attention which is given to the influence of demand on value; and the longer the period, the more important will be the influence of cost of production on value. For the influence of changes in the cost of production takes as a rule a longer time to work itself out than does the influence of changes in demand. The actual value at any time, the market value as it is often called, is often more influenced by passing events whose action is fitful and short-lived, than by those which work persistently. But in long periods these fitful and irregular causes in large measure efface one another's influence; so that in the long run persistent causes dominate value completely.

Thus the value of a forest is not WTP or even WTA, but the social opportunity cost of the forest: what it will mean to society if the forest were not there. This will translate into a larger concentration of carbon dioxide, etc., in the atmosphere, increased soil erosion, and loss of plant and animal biodiversity, etc. Therefore, the value of a forest cannot be measured by the willingness to pay (for visits to the forest), or even the willingness to accept payment for the *loss* of the forest (WTA) to *individual consumers*. The point is that the consumer has little or no information on the vast ecological functions that forests fulfil (see, for example, Rolston and Coufal, 1991). Thus again the neoclassical valuation criterion is flawed, no matter whether it is WTP or WTA. It is the narrowness of the concept of value in conventional economics, which troubles many philosophers and environmentalists (see Rolston, 1985; Rolston, 1994).

3. The Appropriate Benchmark and Exhaustible Resources

In neoclassical theory, the appropriate benchmark for everything is the *pattern of resource utilization that would occur if there were perfect competi-*

tion. In a competitive equilibrium due to perfect competition, nobody can be made better off. The neoclassical theory of exhaustible resources is also similarly flawed, because it is developed (by Hotelling 1931, and others) on the assumption of perfect competition and private property in the exhaustible resource. But here again the actions of private owners are governed by neoclassical value. When the neoclassical model is extended over time, then the *value of time* itself is revealed by the rate of interest. In a competitive equilibrium, the market rate of interest and social rate of time preference (again this is an individualist conception) are the same. Now this value of time determines the "socially optimal" rate of extraction of the exhaustible resource (see Dore, 1994), and all departures[12] from this extraction path are considered undesirable. But even reversing the time-tilt by appropriate taxes (Dore and Harcourt, 1986) would not slowdown over-exploitation of resources, as long as these very benchmark paths may be grossly unjust to future generations.

To sum up, this part: neoclassical theory also determines the value of time in the form of the competitive rate of interest. This rate of interest is also based on consumers' willingness to "trade" goods over time, i.e., defer consumption. This rate of interest is thus also a product of WTP, over time. This then becomes *the benchmark* for judging the extraction rates from a so-called "social viewpoint" in the neoclassical theory. But this is not really a social viewpoint; it is individualistic and utilitarian, which ignores the rights of future generations. Neoclassical theory simply defines it as a "social viewpoint." As the theory of exhaustible resources *also* depends on a willingness to pay measure, it follows logically that it is also flawed.

Thus, in summary, it has been shown that all the arguments point to willingness to pay (or its minor variant, the willingness to accept) as the dominant criterion of valuation; both are based on consumptionist demand. The concept of value based on it applies to goods, and it also applies to the neoclassical valuation of time. If follows that it also applies to the neoclassical theory of the extraction of exhaustible resources. It is this criterion of valuation that is flawed, and it leaves the whole of neoclassical theory deeply flawed in dealing with the analysis of environmental issues.

The objective of the above discussion was not to deny the validity of the critique[13] of neoclassical theory, but to draw attention to a fundamental shortcoming of it, namely, the criterion of value, which is an integral part of the theory. This theory and its valuation criterion rest on the Walrasian model of general equilibrium, which is a model of pure exchange of initially given goods, with no production. Here "value" emerges from the actions of free and voluntary traders, who are aided in a costless fashion by a mythical auctioneer. From the perspective of environmental ethics, it is this criterion of value that is of fundamental importance, rather than the details of the simplistic assumptions of neoclassical demand theory.

Can Timber Production Be Compatible with Conservation of Forest Biodiversity?– Two Case Studies of Plant Biodiversity in Managed Neotropical Forests

Florencia Montagnini
Bryan Finegan
Diego Delgado
Beatriz Eibl
Lilian Szczipanski
Nelson Zamora

SUMMARY. A variety of practices can greatly diminish the impacts of forest management on biodiversity. A case study from a subtropical humid forest reserve in Misiones, Argentina is presented to illustrate a "uniform spacing" method of forest harvest. In this method, trees are selected for extraction or marked for retention according to their scarcity,

Florencia Montagnini, Bryan Finegan, and Diego Delgado are affiliated with the Management and Conservation of Tropical Forests Area, Tropical Agricultural Research and Higher Education Center (CATIE), 7170 Turrialba, Costa Rica (E-mail: montagni@catie.ac.cr).

Beatriz Eibl and Lilian Szczipanski are affiliated with Facultad de Ciencias Forestales, Universidad Nacional de Misiones, (3382) Eldorado, Misiones, Argentina.

Nelson Zamora is affiliated with Instituto de Biodiversidad (INBIO), San José, Costa Rica.

Paper presented at the International Conference: "The Human Dimensions of Global Climate Change and Sustainable Forest Management in the Americas," National University of Brasilia, Brasilia, Brazil, Decebmer 1-3, 1997.

[Haworth co-indexing entry note]: "Can Timber Production Be Compatible with Conservation of Forest Biodiversity?–Two Case Studies of Plant Biodiversity in Managed Neotropical Forests." Montagnini, Florencia et al. Co-published simultaneously in *Journal of Sustainable Forestry* (Food Products Press, an imprint of The Haworth Press, Inc.) Vol. 12, No. 1/2, 2001, pp. 37-60; and: *Climate Change and Forest Management in the Western Hemisphere* (ed: Mohammed H. I. Dore) Food Products Press, an imprint of The Haworth Press, Inc., 2001, pp. 37-60. Single or multiple copies of this article are available for a fee from The Haworth Document Delivery Service [1-800-342-9678, 9:00 a.m. - 5:00 p.m. (EST). E-mail address: getinfo@haworthpressinc.com].

their horizontal distribution, and their characteristics to serve as seed trees. Three years after harvesting, the forest cut by uniform spacing had the highest density of seedlings of total as well as commercial species, and it also exhibited high diversity of understory plants other than trees.

A case study of a silvicultural experiment carried out by CATIE in Costa Rica is also presented, focusing on the effects of logging and post-harvest silvicultural treatments on forest species richness and composition during the first 6-7 yr following logging and 5 yr following the application of treatments. The forest studied was shown to exhibit marked compositional variation in relation to a topographical gradient after the implementation of the experiment; such β- or ecosystem diversity should be taken into account in all future evaluations of the effect of forest management on plant biodiversity. Post-harvest silvicultural treatments caused an immediate reduction in species-richness $\geq$ 10 cm dbh in 1.0 ha due to the chance elimination of species represented by one or a small number of individuals in the plots, and may affect overall species composition in the medium term as they are directed to favoring commercial species by decreasing competition from non-commercial species. Species richness declined under both silvicultural treatments, however, no changes of species-richness or composition as a result of silvicultural treatment were evident in the forest understory (individuals 2.5-9.9 cm dbh). The only detectable changes in understory plant biodiversity were caused by the felling and extraction of timber, and these changes were restricted to the very localized areas disturbed by these management operations.

The application of low-intensity logging methods with selection of remnant trees requires detailed pre- and post-harvest forest inventories and follow up of treatments to ensure successful regeneration and growth of desired species. The Costa Rican study shows that different forest management operations (logging and silviculture in this case) may have different effects on plant biodiversity; this perspective should be applied to future studies in order to better understand the balance between production and conservation in tropical forests. Long-term follow up is needed to assess the true effects of forest management on biodiversity, including natural regeneration as well as growth dynamics of trees, until a whole harvest cycle is completed. *[Article copies available for a fee from The Haworth Document Delivery Service: 1-800-342-9678. E-mail address: <getinfo@haworthpressinc.com> Website: <http://www.haworthpressinc.com>]*

KEYWORDS. Biodiversity, logging, regeneration, shelterwood, silviculture, uniform spacing

INTRODUCTION

Many tropical and subtropical humid forests have been subject to human intervention for centuries, but with the low intensity of management usually employed, environmental functions and species diversity were generally well preserved (Gómez-Pompa 1991). In contrast, most current management schemes for the production of timber result in much greater physical disruption of the forests. However, a variety of practices can greatly diminish the impacts of forest management on biodiversity. For example, it has been suggested that selective logging should strive for minimal canopy opening and minimal soil disturbance (Cheah 1991, Vanclay 1992). Damage to the forest structure can be reduced by decreasing the intensity of timber harvesting and by improving logging practices (Bertault & Sist 1995, Sayer et al. 1995).

Many tropical countries have recently changed forest management regulations to make them compatible with the principle of sustained yield and maintenance of biodiversity (Boyle & Sayer 1995). Specific guidelines are needed to cover a vast array of forest conditions at both large and small scales. These guidelines should be adjusted to suit the scale and objectives of management as well as the light and other resource requirements of the species involved. In addition, methods are needed to evaluate ecological indicators that can serve to verify effects of management on long-term forest productivity and maintenance of biodiversity (Lowe 1995).

In this paper we present examples of two forests experimentally managed for production, located in a subtropical humid region (Misiones, Argentina) and in a tropical lowland humid region (Costa Rica). The management systems used involved reduced impact logging techniques and selection of residual trees. The studies were part of larger programs to design techniques for sustainable forest management throughout the region.

In the Misiones case study, estimation of forest regeneration following timber harvest was used as a measurement of the effects of different harvest methods on forest biodiversity. In Misiones, no silvicultural treatments were applied following timber harvest. Although the studies in Misiones also include the growth dynamics of trees following harvest, results were available only for seedlings and saplings. In the Costa Rican case study described here the criteria used in selecting remnant trees and the timber harvest methods employed were similar to those used in Misiones. However, in contrast to Misiones, post-harvest silvicultural treatments were used to favor the growth of desired trees. Additionally, in this case the evaluation of forest composition included individuals > 10 cm dbh.

Uniform Spacing Method in the Subtropical Forest of Misiones, Argentina

The Province of Misiones, in northeast Argentina (25-28°S, 53-56°W, 100-800 m elevation), has an area of about 30,000 km², less than 1% of the country total (Margalot 1985); however, it produces over 75% of the country's timber (Ministerio de Ecología y Recursos Naturales Renovables 1993). The Misiones forest is part of the Paranaense province of the eastern subtropical forest region (Dimitri 1979). The mixed Paranaense forest formerly covered more than 100,000 km² in regions of Argentina, Paraguay and Brazil, but has been reduced to less than 10% of its original size (Chévez & Gil 1995). The Paranaense forest is one of the most diverse ecosystems of both Argentina and Paraguay (Dimitri 1979). Annual precipitation ranges from 1700 to 2400 mm per year. The mean temperature of the hottest month (January) is 25°C with a maximum of 39°C, and the mean temperature of the coldest month (July) is 14°C with a minimum of −6°C (Eibl et al. 1994). The Guaraní Reserve (26°15' S, 54°15' W, 267-574 m elevation) is a 5,340 ha forest reserve located east of Misiones, near the Brazilian border. The Guaraní forest is part of the 250,000 ha Yabotí Reserve, which was designated a Biosphere Reserve in 1994 (Chévez & Gil 1995). The predominant soils in Guaraní are Ultisols, great group kandiudults (US Soil Taxonomy), regionally known as "red soils" (Laserre 1980). They are deep, clayey, with a predominance of sesquioxide and kaolinite, well drained and well structured, with topsoil organic matter ranging between 3 and 8%, cation exchange capacity of 10-20 cmol/kg, pH in water 5.0-5.5, and base saturation of 50-60%. The relief is rolling hills with slopes ranging from about 5 to 15%.

Under the minimum diameter cutting method prevalent in Argentina, commercial species are extracted if they have a minimum stipulated diameter that ranges from 40 to 55 cm dbh (diameter at breast height) depending on the species (Ministerio de Ecología y Recursos Naturales Renovables 1987). This method was set by the government to make forest cutting economically attractive, but it was established without substantial knowledge of the biology and regeneration dynamics of the species. In the present study, tree regeneration was examined following timber extraction by the conventional minimum diameter method and by a low-intensity, "uniform spacing" method. Regeneration of commercial and non-commercial species following both methods of selective cutting was compared to that of an uncut forest and also to a forest that had been cut by minimum diameters 30 year ago. The life form composition of understory plants other than trees and the degree of canopy closure were also examined.

METHODS

Experimental logging methods–Logging intensity in Guaraní was based on results of a pre-felling inventory that was described by Grance and Maiocco (1990, 1993). For the forest cut by minimum diameters, cutting intensity was 39% of the commercial basal area. The impacts of extraction using this method were particularly detrimental for species that grow in patches, such as *Ocotea puberula*, whose density of individuals ranged from 0 to 30 trees/ha. In one of the eight plots censused, 90% of *Ocotea puberula* trees had diameters at breast height that were larger than the minimum, thus when using the minimum diameter cutting method, such areas of forest were almost cleared of mature individuals of this species that could serve as seed sources (Grance & Maiocco 1993).

In contrast, in the uniform spacing method, many commercially important trees with diameters large enough to be cut under the diameter limit method were left uncut. In the pre-felling inventory, trees were selected for extraction or marked as residuals for retention according to the following criteria: (1) Species scarcity: exploitation was less intense for the scarcest species; (2) Species horizontal distribution: for the species with a patchy distribution, exploitation was lower in the less dense areas, where only 1-4 trees of commercial size/ha were found; (3) The remaining trees displayed adequate timber characteristics (straight, healthy bole) as well as healthy crowns; (4) The remaining trees were mature individuals, but they did not look too old (i.e., they did not show evidence of senescence: damaged crowns, dead branches or signs of decomposition at the base of the trunk), so that they would produce seeds for a prolonged period of time. By combining these four criteria, a more homogeneous spatial distribution of residual trees was obtained (Grance & Maiocco 1993).

In both cutting methods, logging roads were opened prior to logging, following prescribed specifications to avoid as much damage as possible during mechanized felling and skidding. Lianas and bamboos were cut with a machete about a month before felling. Directional felling was done whenever possible. Under the minimum diameter method, a volume of timber comparable to the average for the region (about 37 m³/ha) was extracted. Use of the uniform spacing method resulted in a cutting intensity of about half that by minimum diameter. Likewise, the uniform spacing method resulted in about half as much damage to the remaining trees than in the adjacent forest cut by diameter limits (Grance & Maiocco 1990, 1993).

Tree regeneration studies–Permanent plots were established at random in each of the three logged forests and the uncut forest. A total of 14, 15, 20 and 12 plots were sampled for each height class in US3, MD3, UC and MD30, respectively. The number of plots in each site is currently being increased to reach 25 plots per site for long-term studies of regeneration. The plots were rectangular, with their size varying according to the categories of seedlings/saplings sampled (Montagnini et al. 1998). Each plot was comprised of four nested subplots. The smallest nested subplots, 1 m × 5 m (5 m²), were used to sample the smallest seedlings (Class I, < 10 cm height). With a set width of 1 m, the subplot length was increased to 15 m to have 15 m² subplots for the intermediate size classes (Class II, 10-49 cm height and Class III, 50-149 cm height). Two 15 m² subplots side by side formed a 2 m × 15 m (30 m²) plot for the saplings (Class IV, 150-299 cm height and Class V, > 300 cm height but < 10 cm dbh).

Seedlings and saplings of commercial and non-commercial species were identified, and their height and number per plot were recorded. The relative density and the relative frequency of commercial species were calculated for each site. Importance values (IV) were calculated as the average of relative density and relative frequency for each commercial species and site. Limitations in site replication at Guaraní, as explained before, resulted in "pseudo-replication" as described by Hurlbert (1984); therefore a conventional analysis of variance could not be used. Instead, means and standard errors were used to graphically compare numbers of tree seedlings among treatments.

Plot sizes and height classes were based on previous research in the same forests: for the smallest height class, 5 m² plots gave similar numbers of seedlings/ha and with a comparable standard deviation to those of 10 m² or 15 m² plots (Eibl et al. 1993). For the intermediate height classes, the number of seedlings/ha leveled off when plot size was 15 m², while the number of saplings only leveled off when using 30 m² plots.

Characterization of the understory (plants other than trees) and canopy closure–Frequency of understory species other than trees was calculated as the proportion of the total number of plots which contained a given type of understory plant (bamboos, ferns, grasses, small shrubs). Canopy cover at the center of each 30 m² plot was classified as open when the distance between tree crowns was greater than the average horizontal crown diameter; medium, if the distance between crowns was less than their average diameter; and closed, when the crowns overlapped.

RESULTS

Tree regeneration in the logged and uncut forests–Comparing the mean number of tree seedlings/saplings of both commercial and non-commercial species, more than twice as many were found in the forest cut by uniform spacing three years ago (US3) (mean = 54,330/ha, standard error = 6870), than in the forest cut by minimum diameter 3 years ago (MD3) (mean = 22,270/ha, s.e. = 4620). In the uncut forest (UC) an intermediate number (mean = 32,830/ha, s.e. = 13,550) seedlings/saplings were found. In the forest cut by the minimum diameter method 30 years ago (MD30) the mean (50,000/ha, s.e. = 11,170) was similar to that found in US3. The greatest differences were found between MD3 and US3, and between MD3 and MD30.

When sorted according to their commercial value, the highest numbers of seedlings of commercial species were again found in US3, and the lowest in UC (Montagnini et al. 1998). In contrast, the highest numbers of seedlings of non-commercial species were found in MD30, and the lowest in MD3. US3 had about three times as many seedlings of non-commercial species as MD3. UC had a similar number to US3. UC and MD30 had the highest total (commercial + non-commercial) species richness with 46 species, followed by MD3, which had 34 species, and by US3, which had 28 species (Montagnini et al. 1998).

The uncut forest had the highest richness of commercial species with 18 species, while MD3 had the least richness in this category with 13 species (Table 1). *Apuleia leiocarpa* was the species of highest

TABLE 1. Importance values of seedlings and saplings (all sizes classes together) of commercial species in the four forests of the Misiones study.

Species	Treatment			
	Uniform spacing (US3)	Minimum diameter (MD3)	Uncut forest (UC)	Minimum diameter (MD30)
Apuleia leiocarpa (Vog.) J. Macbr.	34.8	31.7	14.4	2.9
Balfourodendron riedelianum (Engl.) Engl.	3.5	14.5	6.0	9.4
Bastardiopsis densiflora (Hook et Arn.) Hassl.	–	–	0.8	–
Cabralea canjerana (Vell.) Mart.	5.4	11.2	2.5	10.1
Cedrela fissilis Vell.	1.3	11.4	3.1	7.5
Cordia trichotoma (Vell.) Johnst.	–	–	0.8	–
Didymopanax morototoni (Aubl.) Dec. et Planch.	0.7	0.7	–	–
Erythrina falcata Benth	–	–	0.8	–
Lonchocarpus leucanthus Burk	–	–	6.6	–
Lonchocarpus muehlbergianum Hassl.	–	–	1.0	4.7
Leuhea divaricata Mart.	–	0.8	0.9	–
Myrocarpus frondosus Allemao	1.9	–	2.6	7.7
Nectandra lanceolata Nees et Mart. ex Nees	8.7	8.7	1.1	1.7
Nectandra saligna Nees et Mart. ex Nees	4.1	2.1	41.3	4.5
Ocotea diospirifolia (Meisn.) Mez. Emb. Hass	7.0	9.0	4.7	21.9
Ocotea puberula (Nees et Mart.) Nees	11.4	6.7	1.7	20.8
Parapiptadenia rigida (Benth.) Brenan	9.6	–	6.3	4.8
Patagonula americana L.	0.6	–	2.5	–
Peltophorum dubium (Sprengel) Taub	–	–	–	0.9
Prunus subcoriacea (Chod. Et Hassl.) Koehn	10.8	3.2	3.3	2.1
Ruprechtia laxiflora Meisn.	–	–	–	0.9

relative importance value (IV) in US3 and MD3 (34.8% and 31.7%, respectively) (Table 1), while this species ranked second in UC with 14.4%, and ranked one of the lowest in MD30. The Lauraceae (*Nectandra* spp. and *Ocotea* spp.) was the best represented family in all four sites with almost half of the total IV in UC and MD30, but with a relatively smaller value in both recently cut forests, MD3 and US3. Four commercial species in UC were absent in the other three forests: *Bastardiopsis densiflora, Cordia trichotoma, Erythrina falcata* and *Lonchocarpus leucanthus. Cedrela fissilis* was the only commercial species found in US3 and MD3 that was not found in UC. Two species were only found in MD30 and were absent in the other three forests: these were *Peltophorum dubium* and *Ruprechtia laxiflora*, although they had relatively low IVs. MD30 had a more evenly distributed range of IVs than the other forests, with a maximum value of 21.9% for *Ocotea diospirifolia*.

Ten tree/shrub species of no current commercial value were shared by the four forests (Montagnini et al. 1998). The highest species richness in this category was found in MD30 with 32 species, followed by UC.

Types of understory vegetation and canopy cover–US3 and MD30 had the highest heterogeneity in understory species composition (plants other than trees) (Table 2). In US3, there was a predominance of bamboos but there were also shrubs such as *Solanum* spp. and *Trichilia* spp., and vines such as *Bauhinia* spp. In MD3, four species of bamboo were present in all sampled plots: *Bambusa guadua* Humb. & Bonpl. (tacuaruzú), *Chusquea ramosissima* Lind. (tacuarembó), *Guadua trinii* (Nees) Ruprecht. (yatebó), and *Merostachys clausseni* Munro (tacuapí). The arborescent ferns *Alsophila* spp. and *Dicksonia* spp. were found in 73% of the plots. MD30 was similar to US3 in that there was an almost equal distribution of bamboos, ferns, shrubs and vines, as well as a few grasses, such as *Eragrostis* spp., *Paspalum* spp., *Setaria* spp., *Sporobolus* spp. and *Trichloris* spp. In UC there were only bamboos.

The more heterogeneous understory found in US3 and MD30 was related to a greater assortment of canopy cover conditions, with varying proportions of plots with open, medium and closed cover found in these forests (Table 3). In contrast, the predominance of bamboos in MD3 and UC corresponded to a greater proportion of open plots in these sites.

TABLE 2. Frequency of understory life-forms other than trees in the four forests of the Misiones study. Treatment codes: US3: Spacing method applied three years ago (n = 14), MD3: Minimum diameter method applied three years ago (n = 15), UC: uncut forest (n = 20), MD30: Minimum diameter method applied 30 years ago (n = 12).

Life-form (%)	Treatment			
	US3	MD3	UC	MD30
Bamboos	86	100	100	75
Ferns	44	73	0	67
Shrubs	42	0	0	75
Vines	22	0	0	58
Grasses	0	0	0	17

TABLE 3. Frequency of canopy cover categories in the four forests of the Misiones study. Open: distance between tree crowns > average horizontal crown diameter. Medium: crowns do not touch but distance between crowns < average crown diameter. Closed: crowns touch or overlap. No cover: plot at or near center of forest gap of diameter > average tree height. Treatment codes as in Table 2.

	Treatment			
Canopy cover (%)	US3	MD3	UC	MD30
No cover	0	0	0	10
Open	7	80	67	27
Medium	21	20	17	45
Closed	72	0	16	18

DISCUSSION

Tree regeneration following timber extraction–The forest cut by uniform spacing had the highest density of seedlings and saplings of total as well as commercial species, and it also exhibited high diversity of understory plants other than trees. In contrast, the uncut forest had the lowest density of seedlings of commercial species and a predominance of bamboos in the understory. Due to its relatively higher heterogeneity of canopy cover types and understory life forms (Tables 2 and 3), the forest treated by uniform spacing may have appropriate conditions for growth of a higher number of species than undisturbed forest. Differences in the size of canopy openings created by natural disturbance, or in this case by timber extraction, promote differentiation of regeneration, survival, and growth among tree species (Bazzaz 1990, Brown 1993). Therefore, application of low-intensity logging methods with selection of remnant trees, such as with the uniform spacing system described here, may in the long term result in changes in forest composition, with presumably a larger proportion of desired species than in uncut forests.

It should be pointed out that both the preliminary nature of these results and the lack of replication of the treatments limit further interpretation and extrapolation to other situations. In the long term, species composition of each forest will depend on resulting light conditions from the degree of canopy opening in each method, apart from immigrant seeds from nearby forests. Additionally, seed germination and seedling survival may be affected by different impacts on soil characteristics associated with each logging method. The availability of resources other than light, as well as intra- and inter-specific competition may also influence species distribution and growth.

Applicability of the uniform spacing method in forest reserve management–In other research in the same forests, a separate analysis was done for the commercial species > 150 cm height (Classes IV and V), which have a greater possibility of reaching harvestable size: there were means of 400 saplings/ha in US3, 200 in MD3, 220 in UC, and 390 in MD30 (Montagnini et al. 1995, Eibl et al. 1996). In some methods of selective cutting that include management of natural regeneration such as the tropical shelterwood system, a minimum of approximately 100 saplings/ha of commercial value is expected to ensure a good future timber harvest (Lamprecht 1990). For humid tropical forests on good soils, a minimum safe standard of about 200 saplings/ha of commercial species has been suggested (Bruenig 1996). The values reported by Eibl et al. (1996) suggest that the uniform spacing method can provide an adequate stock of saplings of commercial species for the forests of the Guaraní reserve and other similar forests throughout the region.

Results of previous experiences in the region suggest that competition by aggressive understory plants is a major constraint to growth of regenerating seedlings in these forests (Schultz 1967, Eibl et al. 1993). Bamboo apparently out-competes seedlings for light and physically obstructs growth by forming patches of dense overhanging vegetation. In Peninsular Malaysia forest management guidelines indicate that excessive canopy openings should be avoided to prevent the growth of bamboo (*Dendrocalamus* spp.; Cheah 1991).

The minimum diameter method is currently the predominant cutting method because of the larger extractable volume and consequently greater short-term financial benefits it renders. However, the uniform spacing method may be more economically sensible in the long term because it may allow for an earlier second harvest. In evaluating the two methods, consideration should be given to the trade offs between financial profits and the environmental benefits associated with maintenance of species diversity. The uniform spacing method uses detailed guidelines for selection of remaining trees, and requires careful pre-and post-harvest evaluations. The uniform spacing method should be applicable in the small farms in the Yabotí Biosphere Reserve and other forests in the region. In spite of their preliminary nature, the results suggest that if the minimum diameter method is modified to decrease harvest intensity and to include selection of residual trees, it also could contribute to maintaining tree species diversity in these forests.

A CASE STUDY IN THE ATLANTIC LOWLANDS OF COSTA RICA

As part of CATIE's ongoing research on the effects of forest management on forest biodiversity, the present paper reports initial results concerning the effects of logging and silvicultural treatment on species richness and understory composition in a managed Central American lowland rain forest; reports on the early response of forest dynamics to management operations, including stand structure and diameter increment patterns among 106 tree species, are given by Finegan and Camacho (1999) and Finegan et al. (1999), and give essential background information to the present paper.

The study is in progress at Tirimbina Rain Forest Center (henceforward referred to as "La Tirimbina"), near La Virgin, Sarapiquí Canton, Heredia Province, Costa Rica (10° 24′ N, 84° 06′ W) at an altitude of 180-200 m above sea level on the foothills of the Central Volcanic mountain range of the country. The life zone is tropical wet forest (Holdridge's system) with mean annual precipitation near 4,000 mm and mean annual temperature 24.5°C. La Tirimbina lies on highly weathered Quaternary lava flows in which the dominant material is andesitic basalt. A topography of low hills dissected by streams has formed, and the relatively deep, well-drained and infertile clay soils are Ultisols, with pH ca. 4.0 and acid saturation usually > 80%. A full site description is given by Finegan and Camacho (1999) and a full report on the study of plant biodiversity for the 1994-2000 period is in preparation (Finegan et al., in prep.).

METHODS

Forest harvest–The study is under way in a 540 m × 540 m (29.2 ha) experimental plot, managed for the sustainable production of timber in 1989. Timber was harvested from the whole area under strict planning and control in 1989 and 1990 (Quirós & Finegan 1994). The total harvestable volume was 23 m^3/ha but actual harvest intensity was 42% of this (10.1 m^3/ha).

Post-harvest silvicultural treatments–Three different regimes of post-harvest silvicultural intervention were applied to nine 180 m × 180 m (3.24 ha) plots using a complete randomized block design (three replicate plots per silvicultural regime) in 1991 and 1992; these treatments are fully described by Finegan and Camacho (1999). The first intervention regime was the timber harvest already described with no subsequent alterations to the stand, and is subsequently referred to as the control. The other two regimes of intervention involved post-harvest silvicultural treatment. The first treatment was applied by poison-girdling in 1991 and involved a refinement, eliminating most non-commercial trees ≥ 40 cm dbh (diameter at breast height, 1.3 m) and a subsequent liberation of individuals determined to be potential crop trees (only individuals ≥ 10 cm dbh were liberated). Trees deemed to be competing with potential crop trees (see Finegan & Camacho 1999) were poison-girdled only if their dbh was equal to or greater than that of the potential crop tree. The second silvicultural treatment was applied in 1992 and consisted of the formation of a shelterwood, maintaining a continuous high canopy while creating conditions for the regeneration of the more light-demanding commercial species, through the thinning of the middle stories of the forest (carried out by felling by chainsaw).

The effects of silvicultural treatment on forest biodiversity–Forest response to timber harvest and silvicultural treatment (all individuals ≥ 10 cm dbh, including palms but excluding lianas) is being monitored in a square 1.0 ha permanent sample plot (PSP) in the center of each 3.24 ha treatment plot. As silvicultural treatments were applied to each complete 3.24 ha plot, the PSPs were separated from neighboring treatment plots, as well as adjacent unmanaged forest, by 40 m wide buffer strips. Each PSP was divided into square subplots of 20 m × 20 m. Enumeration of the complete set of nine PSPs was implemented in 1990 and measurements were subsequently carried out at 1-2 yr intervals. In order to be able to investigate possible relationships of forest composition to substrate variation, each 20 m × 20 m sub-plots was scored according to its location in one of three topographic positions: hilltops, slope and valley bottoms.

During 1994-1996, a total of 80 sub-plots of 5 m × 5 m were established at random within each of the six PSPs assigned to either the control or liberation/refinement treatments, giving a total area of 0.2 ha within each 1.0 ha PSP. All individuals ≥ 2.5 cm and < 10 cm dbh were monitored in these sub-plots, this size-class being henceforth referred to as the understory, and the procedures already described used for scoring each plot in relation to its topographical position. The 5 m × 5 m subplots were also scored in relation to their location in the understory habitat mosaic: habitat types considered were understory (sites with no evident disturbance either from logging operations or from natural tree or branch falls); gap (opening in the forest canopy, extending through all canopy layers down to a mean vegetation height of 2 m (Brokaw, 1982); road (3-5 m wide strips levelled by bulldozer during logging operations and used for the transport of logs; as the 5 m × 5 m sample plots used were wider than the extraction roads at La Tirimbina, plots were classified as roads when > 50% of their area fell in this habitat type) and road edge (microhabitats crossing the road-forest understory interface, often including substrate material piled up during road construction). As the third replicate per treatment of the understory sample was not installed until 1996, the formal between-treatment comparisons of species richness presented in this paper are for that year; comparisons between categories of the understory habitat mosaic are for 1994.

Throughout the study, all live individuals in both size-classes were identified by Nelson Zamora and voucher samples were kept in a reference herbarium at CATIE.

RESULTS

Vegetation-substrate relationships–The marked topographical gradient of the La Tirimbina forest appears to play a significant role in the structuring of the vegetation (Finegan et al., in prep.). Table 4 compares the composition of hilltop and valley bottom vegetation in both size-classes sampled on the basis of the importance values (IVs) of the six most important species. The total areas sampled per topographical category in the PSPs were not equal, but the calculations of the IV were done using equal sample areas per category, these being 2.32 ha for individuals ≥ 10 cm dbh and 0.2425 ha for the understory. For each size-class, subplots were randomly selected for analysis in the

TABLE 4. Variation in forest composition related to topographical variation in a managed neotropical lowland rain forest, La Tirimbina, northeastern Costa Rica. The Table gives the importance value (the sum of the relative abundance, basal area and frequency of each species, converted to %) of the six most important species for, (a) individuals ≥ 10 cm dbh, total area sampled 2.32 ha in each topographical position, and (b) the understory, individuals ≥ 2.5 cm dbh and < 10 cm dbh, total area sampled 0.2425 ha in each topographical position. The vertical distances separating sites in two categories are approximately 20 m. Data from Finegan et al. (in prep.).

		Hilltop	I.V.	Valley bottom	I.V.
a) ≥ 10 cm dbh		*Pentaclethra macroloba*	14.9	*Pentaclethra macroloba*	20.3
		Ferdinandusa panamensis	6.2	*Iriartea deltoidea*	3.7
		Euterpe precatoria	3.5	*Welfia georgii*	3.4
		Tapirira guianensis	3.4	*Casearia arborea*	2.9
		Welfia georgii	3.3	*Apeiba membranacea*	2.5
		Protium ravenii	3.2	*Carapa guianensis*	2.3
		Subtotal (top six spp.)	34.5	Subtotal (top six spp.)	34.9
		Subtotal (136 other spp.)	65.5	Subtotal (145 other spp.)	65.1
		Total (142 spp.)	100	Total (151 spp.)	100
b) 2.5-9.9 cm dbh		*Euterpe precatoria*	6.8	*Prestoea decurrens*	5.2
		Ferdinandusa panamensis	6.6	*Psychotria luxurians*	4.1
		Geonoma congesta	5.6	*Psychotria elata*	4.0
		Licaria sarapiquensis	4.1	*Warsewiczia coccinea*	3.5
		Protium pittieri	3.7	*Pentaclethra macroloba*	3.4
		Protium ravenii	3.4	*Laetia procera*	2.6
		Subtotal (top six spp.)	23.4	Subtotal (top six spp.)	22.8
		Subtotal (140 other spp.)	76.6	Subtotal (108 other spp.)	77.2
		Total (146 spp.)	100	Total (114 spp.)	100

topographical category of greater area; the total area sampled was used for that of lesser area. Only 5 m × 5 m subplots considered free of direct structural disturbance by timber harvest and silvicultural operations were considered for the analysis of understory composition.

For individuals ≥ 10 cm dbh, two widely-distributed and abundant species, the dominant tree *Pentaclethra macroloba* and the middle-story palm *Welfia georgii*, were among the top six species in both topographical categories (Table 4a). Otherwise, the most important species differ completely between hilltops and valley bottoms–the palm *Euterpe precatoria* is a typical member of the hilltop flora and is almost absent from valley bottoms, for example (see also Clark et al. 1995), while the palm *Iriartea deltoidei* and the canopy tree species *Apeiba membranacea* and *Carapa guianensis* are typical associates of *Pentaclethra* and *Welfia* in valley bottoms; *Apeiba* and *Carapa* are common or abundant components of primary swamp forest on relatively fertile alluvial soils in Central America's Caribbean lowlands (e.g., Webb & Peralta 1998) and it seems possible that they are outcompeted by canopy species such as *Tapirira guianensis* on infertile, well-drained hilltop soils at sites such as La Tirimbina.

There was even less similarity between hilltops and valley bottoms in terms of the most important species of the understory (Table 4b). In contrast to the situation for individuals ≥ 10 cm dbh, no species was among the six most important in both topographical categories. Two closely-related palms were the most important species: *Euterpe precatoria* on hilltops and *Prestoea decurrens* in valley bottoms (see also Clark et al. 1995). Otherwise, the typical species of the understory of hilltops were the colonial palm *Geonoma congesta* and the trees *Ferdinandusa panamensis* (juveniles of this middle story species found in the 2.5-9.9 cm dbh range are probably all reproductively immature) and *Licaria sarapiquensis* (a true understory species, reaching reproductive maturity in this forest stratum), while the rubiaceous species *Psychotria elata*, *P. luxurians* (both true understory species) and *Warsewiczia coccinea* (middle story as adults) characterized the understory of valley bottoms.

The effects of management operations on species richness and composition of the vegetation–An overall total of 256 species of trees and palms, ≥ 10 cm dbh in 9.0 ha and representing 152 genera and 92 families, was recorded during the eight year period reported in this paper (1988-1996; a full species list of trees recorded in the La Tirimbina plots in the period to 1996 is provided by Zamora et al. 1997). At the time of writing, 51 of the tree species identified were considered to be commercial in the Costa Rican national market for timber and plywood.

The three different regimes of intervention brought about marked between-treatment differences in stand structure in this silvicultural experiment; mean stand density in 1996 was 533 trees ha^{-1}, ≥ 10 cm dbh, in the control plots (s.d. 94 trees) and 418 trees ha^{-1} (s.d. 64) in the liberation/refinement plots, for example; the corresponding figures for whole-stand basal area were 23.7 m^2 ha^{-1} (s.d. 1.1) and 18.8 m^2 ha^{-1} (s.d. 1.4) (Finegan & Camacho 1999). Values of species richness ≥ 10 cm dbh in 1.0 ha recorded in individual PSPs during 1990-1996 varied between 81 and 118, with both the lowest and the highest values being found in control plots. Such variability within treatments meant that formal statistical comparisons did not yield significant differences between them. Species turnover (gains and losses) over time, on the other hand, showed unequivocal evidence of between-treatment differences. Species richness declined under both silvicultural treatments, however. Although the largest declines of species richness were recorded in the two shelterwood plots, this parameter changed little in the third PSP of this treatment. There were no interpretable between-treatment differences of vegetation composition for the vegetation ≥ 10 cm dbh during the study period.

There were no differences between post-harvest regimes of silvicultural treatment in means of either numbers of individuals or numbers of species in 0.2 ha in the understory in 1996 (ANOVA, $P > 0.1$), mean values being 432 individuals (s.d. 79) and 156 species (s.d. 7) in the control plots, and 454 individuals (s.d. 79) and 148 species (s.d. 12) in the liberation/refinement plots. In contrast, the timber harvesting operations carried out during 1989-1990 had a clear influence on the understory vegetation, there being marked variation of numbers of individuals and species-richness in 5 m × 5 m plots among components of the understory habitat mosaic in 1994. The main aspect of this variation was the difference between plots in undisturbed patches and those plots located in roads and gaps, in which the numbers of both individuals and species were significantly lower in 1994; plots scored as road edges showed significantly greater numbers of individuals than gaps in that year, but these two habitat types did not differ in their species richness (Tukey tests on rank-transformed data, $\alpha = 0.05$; see Figure 1). Czekanowski similarity coefficients showed that vegetation composition, as well as stem density and species richness in 5 m × 5 m plots, also varied between understory habitats (Table 5). The lowest similarity was evident between the vegetation of roads and that of gaps and undisturbed forest patches, though similarity between the latter habitats was intermediate; road edges were similar to both roads and undisturbed patches as they included elements of the vegetation of both the latter habitats. These results may be interpreted as indicating

FIGURE 1. Median values of number of species (open bars) and numbers of individuals (shaded bars) in 5 m × 5 m plots in different understory habitat types in the managed forest at La Tirimbina, 1994. Upper bars show the maximum value and lower bars, the minimum value; habitats with the same letters above the bars were not significantly different (Tukey test on rank-transformed data, α = 0.05). Adapted from Delgado et al. (1997), *n* per treatment as in Table 5.

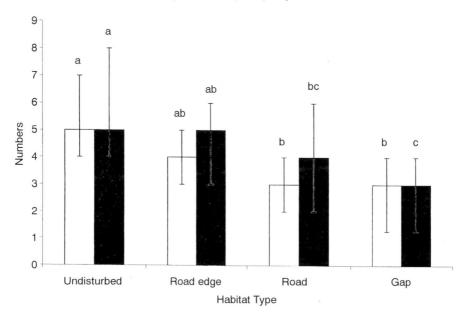

that, as may be expected (cf. Guariguata & Dupuy 1997), road vegetation is dissimilar to that of the rest of the forest habitat. It appears worth emphasizing that road and gap vegetation, though both represent regeneration after drastic disturbances of the vegetation right down to the forest floor, have different characteristics. This latter point is reinforced by values of the Shannon-Weiner diversity index *H'*, calculated for the same number of plots per habitat (18, a total area of 450 m^2): *H'* was > 5.2 for all habitats except roads, in which its value fell to 4.49 (Finegan et al., in prep.).

DISCUSSION

The preliminary results of the study of tree species biodiversity (individuals ≥ 10 cm dbh) in the managed forest at La Tirimbina

TABLE 5. Values of Czekanowski's coefficient of similarity between the understory vegetation of different components of the habitat mosaic at La Tirimbina, northeastern Costa Rica, in 1994 (timber harvesting operations were carried out during 1989-1990). For each comparison, an equal number of 5 m × 5 m plots per habitat was used; as the numbers of plots per habitat varied widely between habitat types (undisturbed, n = 230, road edge, n = 43, road, n = 29, gap, n = 18), the calcualtion of the coefficient for a given pair of habitats was carried out using all plots for the type with the least number, and the same number of plots randomly selected from those located in the type with the greater number (data from Delgado et al. 1997).

Habitats	Road	Road edge	Gap	Undisturbed
Road	--	0.39	0.11	0.22
Road edge		--	0.26	0.43
Gap			--	0.30
Undisturbed				--

show that it is particularly important to understand the causes and nature of species loss from permanent sample plots in such studies, in order to correctly identify the implications of these results regarding the broader relationship between silvicultural treatment and tree species diversity in tropical rain forests. There were no statistically significant effects of silivcultural treatment on species richness 5 years after treatment. Species richness in 1.0 ha was reduced by silvicultural treatment at La Tirimbina, as a direct result of the elimination of trees which were competing with potential crop trees (the liberation/refinement treatment) or which were shading potential microsites for the regeneration of commercial species (the shelterwood treatment). This result may be understood as follows. Species-abundance distributions of rain forest tree species in PSPs are dominated by species which are "rare" in the plot, i.e., those represented by few individuals, in many cases only one. In a given PSP, a proportion of the individuals killed by relatively drastic disturbances, such as those associated with timber harvesting and silvicultural treatment, are bound to be the only representatives of their species in that plot: thus species richness declines in the plot. This point may be illustrated for La Tirimbina as follows. In 1996, a total of 64 species had been recorded as lost from one or other of the nine PSPs at the site, due to all causes, since the beginning of the study. Fifty-two of those species were represented by only one individual in the plot from which they were lost, nine by two individuals and only three by > 3 individuals. The loss of species from individual plots does not necessarily mean that those species are in overall decline in the forest as a whole; at La Tirimbina, for example, only twelve species of the 64 species were lost from more than one PSP. This is due at least partially to the low likelihood that, in the presence of ca. 100 species ha $^{-1}$, silvicultural treatment will eliminate the same species from more than one PSP. It must be emphasized that where refinement treatments are applied, however, as is proposed, for example, in the CELOS silvicultural system (de Graaf et al. 1996), populations of some common non-commercial tree species may be severely impacted (Finegan et al., unpublished data).

The results of this study show that over the short term, the effects of different forest management operations on plant biodiversity may be sharply delimited. The type of silvicultural treatment applied at La Tirimbina eliminates trees ≥ 10 cm dbh and its effects on species richness could only be detected in that size class during the first five years after application. In the understory, neither stem density nor species richness showed any detectable response to silvicultural treatment during the same period. The felling and extraction of timber, which disturbs the understory directly, however, showed marked though localized effects on both these parameters of the understory vegetation. Finally, it is worth emphasizing that the characteristics of the vegetation which is regenerating post-harvest in extraction roads at La Tirimbina is very different from that of felling gaps, which shows much greater similarity to that of patches undisturbed by the timber harvest. We hypothesize that under the timber harvest intensity applied at La Tirimbina, regeneration in gaps will maintain forest characteristics similar to those of undisturbed forests, while noting that on the other hand, Guariguata and Dupuy (1997) suggest that the vegetation of logging roads in forests similar to La Tirimbina would take > 80 years to recover characteristics similar to those of the surrounding vegetation.

Finally, our analysis of vegetation structure and composition at La Tirimbina revealed a degree of natural heterogeneity, linked to topographical variation, unsuspected when the silvicultural experiment at the site was set up during 1988-1990. Future evaluations of the effects of forest management operations on plant biodiversity should take into account the likely existence of such heterogeneity and be stratified by vegetation type–in the case of La Tirimbina, the strata would be hill-top, slope and valley bottoms.

CONCLUSIONS

- The use of the Uniform Spacing method of forest harvesting can contribute to maintaining biodiversity in tropical forests managed for timber production.
- Application of silvicultural treatments following harvest may immediately reduce species richness in permanent sample plots, by eliminating from the plots, essentially by chance, species represented by only one or two individuals in the size classes covered by the permanent sample plot procedures used; this does not mean that certain species are in overall decline.
- The short-term effects of forest management operations on plant biodiversity depend on the exact nature of the operation, and full understanding of biodiversity changes in managed forests will probably only be achieved through the wider application of sampling protocols or experimental designs which permit the individual assessment of each operation.
- Marked local-scale vegetational heterogeneity is likely at sites where environmental heterogeneity is marked, such as La Tirimbina, and sampling or experimental treatments related to the effects of management on plant biodiversity must be stratified by vegetation type.
- Long-term follow up is clearly needed, however, to assess the true effects of forest management on plant biodiversity.

REFERENCES

Alder, D. and T. J. Synnott. 1992. Permanent Sample Plot Techniques for Mixed Tropical Forest. University of Oxford, Oxford, GB. 124 p.

Bazzaz, F. A. 1990. Regeneration of tropical forests: physiological responses of pioneer and secondary species. *In* pp. 91-118. A. Gómez-Pompa, T. C. Whitmore, and M. Hadley (Eds.). Rainforest Regeneration and Management. MAB Series, UNESCO, Paris.

Bertault, J. C. and P. Sist. 1995. Impact de l'exploitation en forêt naturelle. Bois et Forêts des Tropiques 245: 5-20.

Boyle, T. J. B. and J. A. Sayer. 1995. Measuring, monitoring and conserving biodiversity in managed tropical forests. Comm. For. Rev. 74: 20-25.

Brokaw, N. V. L. 1982. Treefalls: frequency, timing and consequences. *In* pp. 101-108 Leigh Jr., E. G., Rand, A. S., Windsor, D. M. (Eds.). Ecology of a Tropical Forest: Seasonal Rhythms and Long-Term Changes. Smithsonian Institution Press, Washington, D.C.

Brown, N. 1993. The implications of climate and gap microclimate for seedling growth conditions in a Bornean lowland rain forest. J. Trop. Ecol. 9: 153-168.

Bruenig, E. F. 1996. Conservation and Management of Tropical Rainforests. An Integrated Approach to Sustainability. CAB International, Wallinford, Oxon, UK.

Cheah, L. C. 1991. Letter to the editor. J. Trop. For. Sci. 4: 96-99.

Chévez, J. C. and G. Gil. 1995. Misiones hoy al rescate de la selva. Nuestras Aves (Argentina): 5-9.

Clark, D. A., D. B. Clark, R. Sandoval M., and C. Castro. 1995. Edaphic and human effects on landscape-scale distributions of tropical rain forest palms. Ecology, 76: 2581-2594.

Delgado, D., B. Finegan, N. Zamora and P. Meir. 1997. Efectos del aprovechamiento forestal y el tratamiento silvicultural en un bosque húmedo del noreste de Costa Rica: cambios en la riqueza y composición de la vegetación. CATIE, Turrialba, Costa Rica: Serie Técnica, Informe Técnico No. 298. 43 p.

Dimitri, M. J. 1979. Las áreas argentinas de bosques espontáneos. *In* pp. 6-17 D. Cozzo. Encyclopedia Argentina de Agricultura y Ganadería, 2da. Ed., Tomo II, Ed. ACME, Buenos Aires.

Eibl, B., L. Szczipanski, R. Ríos and N. Vera. 1993. Regeneración de especies forestales nativas de la selva Misionera. *In* pp. 100-122. Instituto Subtropical Investigaciones Forestales (Ed.). VII Jornadas Técnicas: Bosque Nativo: Uso, Manejo y Conservación. UNaM, Facultad Ciencias Forestales, Eldorado, Misiones, Argentina.

Eibl, B., F. Silva, A. Bobadilla, E. Weber and D. Gonseski. 1994. Boletín Meteorológico Aeródromo Eldorado. Período 1985/1994. UNaM, Facultad Ciencias Forestales, Eldorado, Misiones, Argentina.

Eibl, B., F. Montagnini, C. Woodward, L. Szczipanski and R. Ríos. 1996. Evolución de la regeneración natural en dos sistemas de aprovechamiento y bosque nativo no perturbado en la Provincia de Misiones-República Argentina. Yvyraretá (Argentina) 7(7): 63-78.

Finegan, B. and M. Camacho 1999. Stand dynamics in a logged and silviculturally treated Costa Rican rain forest, 1988-1996. For. Ecol. Manage. 121, 177-189.

Finegan, B., M. Camacho and N. Zamora. 1999. Diameter increment patterns among 106 tree species in a logged and silviculturally treated Costa Rican rain forest. For. Ecol. Manage. 12, 159-176.

Finegan B., D. Delgado and N. Zamora. Factors structuring the plant community of a tropical rain forest managed for timber production. In preparation.

Gómez-Pompa, A. 1991. Learning from traditional ecological knowledge: insights from Mayan silviculture. *In* pp. 335-342. A. Gómez-Pompa, T. C. Whitmore, and M. Hadley (Eds.). Rainforest Regeneration and Management. MAB Series, UNESCO, Paris.

Graaf, N. R. de, Koning, D., Spierings, M., 1996. Some conditions and possibilities for successful application of the CELOS Management System. Bos. Newsletter 15(34): 74-83.

Grance, L. and D. Maiocco. 1990. Informe cuatrimestral presentado a la Dirección General de Bosques, Pcia. de Misiones, sobre el Plan de Ordenación de Guaraní. Ministerio de Ecología y Recursos Naturales Renovables Posadas, Misiones, Argentina.

Grance, L. and D. Maiocco. 1993. Comparación de dos criterios de entresaca en el bosque subtropical Misionero. *In* pp. 284-299. Instituto Subtropical Investigaciones Forestales (Ed.). VII Jornadas Técnicas: Bosque Nativo: Uso, Manejo y Conservación. UNaM, Facultad Ciencias Forestales, Eldorado, Misiones, Argentina.

Guariguata M. R. and J. M. Dupuy. 1997. Forest regeneration in abandoned logging roads in lowland Costa Rica. Biotropica 29: 15-28.

Hurlbert, S. T. 1984. Pseudoreplication and the design of ecological field experiments. Ecol. Mono. 54: 187-211.

Lamprecht, H. 1990. Silvicultura en los Trópicos, pp. 129-133. GTZ, Eschborn.

Laserre, S. 1980. Los Suelos Misioneros y su Capacidad de Uso para Plantaciones Forestales. Assoc. Plantad. For. Mis., Bol. No. 10. Posadas, Misiones, Argentina.

Lowe, P. D. 1995. The limits to the use of criteria and indicators for sustainable forest management. Comm. For. Rev. 74: 343-349.

Margalot, J. A. 1985. Geografía de Misiones. Industria Gráfica del Libro, Buenos Aires.

Ministerio de Ecología y Recursos Naturales Renovables. 1987. Diámetros Mínimos de Corta. Decreto 1617/87. Posadas, Misiones, Argentina.

Ministerio de Ecología y Recursos Naturales Renovables. 1993. Censo de la Industria de Transformación Mecánica de la Madera de la Provincia de Misiones, Sept. 1992-Julio 1993. Posadas, Misiones, Argentina.

Montagnini, F., B. Eibl, C. Woodward, L. Szczipanski and R. Ríos. 1995. Natural regeneration under two systems of selective cutting and undisturbed forest in Misiones, Argentina. XXth IUFRO World Congress, Tampere, Finland. 6-8 August 1995. Available from the Internet: http//iufro.boku.ac.at/procee/s107/s10700.htm.

Montagnini, F., B. Eibl, C. Woodward, L. Szczipanski and R. Ríos. 1998. Tree regeneration and species diversity following conventional and uniform spacing methods of selective cutting in a subtropical humid forest reserve. Biotropica 30(3): 349-361.

Quirós, D. and B. Finegan. 1994. Manejo sustentanble de un bosque natural tropical en Costa Rica. Definición de un plan operacional y resultados de su aplicación. Serie Técnica. Informe Técnico No. 225. Colección Silvicultura y Manejo de Bosques Tropicales. CATIE, Turrialba, Costa Rica. 25 p.

Sayer, J. A., P. A. Zuidema and M. H. Rijks. 1995. Managing for biodiversity in humid tropical forests. Comm. For. Rev. 74: 282-287.

Shultz, J. P. 1967. La regeneración natural de la selva mesofítica tropical de Surinam después de su aprovechamiento. Bol. Inst. For. Latinoamer. Invest. Capacit. (Mérida, Venezuela) 23: 3-27.

Synnott, T. J. 1978. A manual of permanent plot procedures for tropical rain forests. Commonwealth Forestry Institute, Tropical Forestry papers. No. 14. 67 p.

Vanclay, J. K. 1992. Species richness and productive forest management. *In* pp. 1-9. F. R. Miller, and K. L. Adam (Eds.). Wise Management of Tropical Forests, Oxford For. Inst., Oxford.

Webb, E. L. and R. Peralta. 1998. Tree community diversity of lowland swamp forest in Northeast Costa Rica, and changes associated with controlled selective logging. Biodiversity and Conservation 7, 565-583.

Zamora, N., M. Artavia, D. Delgado and M. Camacho. 1997. Especies vegetales de un bosque tropical húmedo primario manejado, Finca Tirimbina, noreste de Costa Rica. CATIE, Turrialba, Costa Rica: Manejo Forestal Tropical No. 1. 8 p.

Joint Implementation in Costa Rica:
A Case Study at the Community Level

Olman Segura
Klaus Lindegaard

SUMMARY. The policy of joint implementation is emerging as a new strategy for implementing global environmental aims, especially with regard to regulating the climate change process, where emission source and sink countries agree to develop a joint program upon a mixed argument of partnership and cost-effectiveness. Pros and cons have emerged during the development of this system.

Costa Rica is the first country, together with Norway, to launch such a program jointly, and Costa Rica is also the first country developing Carbon Tradable Offset bonds to be sold on the world market as a new commodity. It is hoped that this initiative will help the country and its inhabitants to create better living conditions and economic growth; however, this new institutional transformation and international acceptance of this new instrument are only just beginning to develop.

This, therefore, provides a very interesting field for research from a distinct perspective. We chose to start searching for positive or negative impacts at the community level. In this sense the paper deals with questions such as: What happens at the community level?; Is it possible

Olman Segura is affiliated with Centro Internacional en Politica Economica (CINPE), Universidad Nacional, Costa Rica.

Klaus Lindegaard is affiliated with the Department of Business, Aalborg University, Denmark.

Paper presented at the Seminar: The Human Dimensions of Global Climate Change and Sustainable Forest Management in the Americas. Brasilia, Brazil, December 1-3, 1997. The authors thank comments from Adam Drucker. The usual disclaimers apply.

[Haworth co-indexing entry note]: "Joint Implementation in Costa Rica: A Case Study at the Community Level." Segura, Olman, and Klaus Lindegaard. Co-published simultaneously in *Journal of Sustainable Forestry* (Food Products Press, an imprint of The Haworth Press, Inc.) Vol. 12, No. 1/2, 2001, pp. 61-78; and: *Climate Change and Forest Management in the Western Hemisphere* (ed: Mohammed H. I. Dore) Food Products Press, an imprint of The Haworth Press, Inc., 2001, pp. 61-78. Single or multiple copies of this article are available for a fee from The Haworth Document Delivery Service [1-800-342-9678, 9:00 a.m. - 5:00 p.m. (EST). E-mail address: getinfo@haworthpressinc.com].

61

to realize joint implementation with positive local, social and economic impacts?; and What are the necessary conditions for this to become successful? *[Article copies available for a fee from The Haworth Document Delivery Service: 1-800-342-9678. E-mail address: <getinfo@haworthpressinc. com> Website: <http://www.haworthpressinc.com>]*

KEYWORDS. Joint implementation, carbon sequestration, community, secondary forest

JOINT IMPLEMENTATION

The policy of joint implementation is emerging as a new strategy for implementing global environmental aims, especially with regard to regulating the climate change process, where funds from rich emissions source countries are allocated directly to projects in other (poor) countries based upon a mixed argument of partnership and cost-effectiveness. There are in principle different types of projects. These include either allocating funds for the solution of environmental problems in poorer countries or allocating funds to poorer countries as a means of remedial or compensatory action against ones own problems.

The traditional strategy towards international environmental action is based mainly on two different programs. First, the special programs of the international organizations supporting projects in the poor countries with the financial means allocated from the richer countries, or second, on international agreements between emission source countries, which are implemented nationally sometimes with some special support schemes or arrangements for the poorer countries.

The absence of supranational authorities with the power to implement the polluter pays principle (PPP) through pollution taxes and marketable pollution permits together with the general lack of economic incentives in the international allocation of abatement effort, has spurred the development of bilateral bargaining solutions and market creation (Zylicz 1991). Standard economic analysis points here to the decisive role played by transaction and negotiation costs in explaining the patterns of bilateral bargaining solutions to joint and global environmental problems. This standard approach to dispute settlement between parties joining international environmental conventions underlines the role of the question of national sovereignty, opportunism and power. Dispute settlement builds first of all on dispute avoidance via monitoring, reporting and inspection, then on well specified non-compliance procedures, followed by consultation and negotiation, mediation, conciliation, arbitration and, finally judicial settlement in the international Court of Justice with its special environmental branch. However, all these measures depend on the voluntary participation of the parties (nation states) involved (OECD, 1995). Hence, it is possible for nations not only to avoid binding international agreements but also to engage in prolonged disputes over the actual interpretation of international agreements, if they choose to do so.

On the other hand, we see a growing number of international agreements together with a growing understanding of the global and common nature of many of the environmental problems, that earlier were considered as purely local and exclusively national. In this way, we will regard in principle all kinds of cross-national bargaining solutions, whether multilateral or bilateral, as joint implementation.

Implementation of cross-national/global environmental goals have developed into the following two main categories and associated subcategories:

A. Different Types of International Joint Implementation Projects

1. International aid and support programs of funds for solution of environmental problems in poor countries (e.g., UN programs)
2. International agreements to solve common environmental problems (e.g., Vienna/Montreal CFC treaties)

B. Different Types of Bilateral Joint Implementation Projects

1. State funds for solution of environmental problems in other countries (e.g., Western European countries investing in Eastern European countries, Northern European countries investing in Southern European countries)
2. State funds for remedial or compensatory action of their own environmental problems in other countries (e.g., carbon sink projects)
3. State funds for a mix of 1 and 2 above (e.g., dept-for-nature swaps)

4. Private funds for solution of 1, 2 or 3 above (e.g., environmentalist group financed conservation projects)

In relation to the combating of global climate change impacts of human activities, it is useful to distinguish between different types of strategies and projects. Greenhouse gas emission mitigation can be approached by fossil fuel saving, by substitution of energy sources towards renewable energy, improvements of energy efficiency, changes in industrial technologies and substitution of CFCs, etc., as well as through changes in agricultural practices leading to reduced methane emissions. Furthermore, carbon sink enhancement by changes in land-use and reforestation or forest conservation also plays an important role.

Article 4.2 of the 1992 United Nations Framework Convention on Climate Change recognizes, in principle, joint implementation projects to combat climate change if they meet the prerequisites of actually contributing to the reduction of global greenhouse gas emissions, if this effect is controllable and verifiable, if the costs of achieving the emission targets are lower than purely national investments and if both the investing and the receiving countries are better off implementing the project than not, given the total costs and benefits of the project (Torvanger 1993).

The Kyoto Protocol of December 1997 has now acknowledged and institutionalized the joint implementation strategy towards global climate change. Here the protection and enhancement of greenhouse gas sinks and reservoirs is emphasized both by the promotion of sustainable forest management practices, afforestation and reforestation, and by promoting joint implementation projects and acknowledging emission trading. Emissions of 1990 are taken as the baseline regarding verifiable human-induced land-use change and forestry activities as well as greenhouse gas emissions. The parties to the protocol must include the removal of anthropogenic emissions by sinks in their annual inventory of emissions by sources, which is to be subject to international expert review (United Nations 1997).

POTENTIAL ADVANTAGES AND DISADVANTAGES OF JOINT IMPLEMENTATION

At a global level the projects present the potential advantage of increasing the incentives to reduce greenhouse gas emissions, developing new technologies, encouraging cross-country commitments and reducing the overall costs of implementation of international targets. Donor countries can benefit from cost savings, obtaining a national share of global climate benefits and new potential investment and export markets. For the receiving countries, the advantage should be in terms of access to additional financial resources, transfer of technologies and potential cost savings due to new technology, obtaining a national share of the global climate benefits, obtaining national and local environmental benefits, job creation and capacity building (Selrod et al. 1995).

Some possible disadvantages of joint implementation have also been recognized. The whole question of monitoring, control and verification of the investment projects is very complex, along with the uncertain effects on technological change and abatement efforts in the donor country, together with the possible distortion of development preferences and opportunities in the receiver country, the increased foreign influence over the management of natural resources and the overall global equity effects of the projects.

THE COSTA RICAN CARBON BONDS

The Costa Rican and Norwegian governments very recently achieved an agreement under the joint implementation initiative. Costa Rica issued carbon bonds (Greenhouse Gas Emissions Mitigation Certificates) for a value of 2 million dollars thereby permitting Norway to buy a sequestration service of 200,000 tons of carbon from Costa Rican forests. The sequestration service will be provided over a period of 25 years through reforestation and forest conservation projects in Costa Rica. The agreement between the Costa Rican and Norwegian governments is designated as a pilot project of the joint implementation program of the Climate Change Convention (CCC) and it is estimated that Costa Rica has approximately 400,000 hectares of degraded land, which could be reforested in a similar way (Tico Times, 1997).

The agreement has been accompanied by a new institutional set-up in Costa Rica. A special national office for joint implementation, called the Costa Rican Office on Joint Implementation (OCIC, Oficina Costarricense de Implementacion Conjunta) is in charge of the international negotiations and agreements. The carbon funds resulting from the sales of carbon bonds are transferred to the national forest fund (FONAFIFO, Fondo Nacional de Financiamiento Forestal) which invest in national parks, forest conservation and reforestation projects. The monitoring and control of the projects are in the hands of the National System of Conservation Areas (SINAC) of the Ministry of Environment and Energy (MINAE), together with private sector auditors. Individual landowners can submit an application to the national forest fund for financial support for reforestation or forestry protection and FONAFIFO can, furthermore, make use of the funds to support existing national parks as well, and to compensate landowners who must meet regulations on the use of their properties. The Norwegian bonds are worth 10 dollars per ton of carbon. This carbon price has been calculated on the basis of an estimated average income loss per hectare of 50 dollars a year in agriculture and an estimated annual carbon fixation capacity of woodland of 5 tons per hectare, according to a senior Costarican official.

The Costa Rican carbon agreement is based on a mix of public and private Norwegian funds with US$300,000 coming from a private consortium engaged in hydroelectric projects in the area. The Norwegian government is also engaged in a wider joint implementation project between the two countries which includes direct investments by Norway in the modernization of an existing hydroelectric plant, which also has mitigating effects on greenhouse gas emissions in the region. The Costa Rican forestry project copes in this way with climate change both via carbon sequestration and via watershed maintenance for hydroelectric energy production. This can therefore be considered as a really mixed type of bilateral project, a new type B5 (c.f. the first section), addressing environmental problems in both the donor and the receiver country with both public and private funds.

Joint implementation programs, such as in the case of Costa Rica, also allows the generation of several other activities from the same forest without affecting carbon storage services. These examples include: ecotourism; extraction of minor forest products, such as latex, fruits, wildlife, nuts, etc.; and the use and research of biodiversity. Additionally, each one of these activities may generate multiple income streams, because in order to internationally sell the service of carbon sequestration, there is the need for the services of, for example, cartographers, Geographic Information Systems analysts, insurance companies, foresters, engineers, economists, financial system specialists, and others. An entirely new economic cluster of activities is therefore being created around the emerging new commodity of carbon services which is only just starting to be traded internationally.

In short, this activity of carbon sequestration seems to be an especially interesting alternative for less developed countries, though it is also attractive for developed ones, because it not only creates jobs and increases income but also stops deforestation and/or may increase reforestation.

THE STUDY CASE:
JUNQUILLAL DE SANTA CRUZ

As stated above there are several arguments in favor and against the initiative of joint implementation with carbon sink projects. At the national level, the Costa Rican experience seems thus far to be a good opportunity for the country to show the project benefits, if developed properly and with a good verification component. However, even with such a positive example, several questions remain to be answered. For instance: What are the implications at the community level?; Is it possible to realize joint implementation with positive local, social and economic impacts?; and What are the conditions for this to become successful?

In order to address these questions we visited a project in Junquillal de Santa Cruz. This is a very small community located in the Province of Guanacaste in the Northern part of Costa Rica. Activities in this small town of a few hundred inhabitants have traditionally been related to agriculture and cattle ranching. Currently, unemployment exists due to the fact that agricultural activities have been decreasing. Some large landowners had to abandon their land due to the low international price of meat, the high costs of cattle production and the prolonged dry seasons in these areas. Such land has often been sold to the Institute of Agricultural Development (IDA, Instituto de Desarrollo Agrario), a land-tenure institute, since it was under threat of invasion.

A group of thirty landless families from different parts of the country were grouped together and IDA provided each of them with a small parcel (8-10 hectares) of land in Junquillal, approximately 4 years ago. Almost all the people from this town, as well as the newcomers are farmers. They produce rice, beans, maize and other basic crops, and raise pigs, cattle and other animals for their landlords. In general they also realize these activities for subsistence purposes, as well as, for the newcomers, raising one cow per family, following their gaining title to the plot of land and receiving some help from some of the organizations described below. The community of Junquillal along with the newcomers have also often faced the threat of forest fires on the neighboring 200 hectares of secondary forest, which also threatens their poor wooden houses.

IDA along with a program from Food and Agriculture Organization (FAO) has been working with poor rural communities as part of a program called the "Chorotega Forestry Project" (Proyecto Forestal Chorotega). This project provides technical and logistic support to 15 small communities in the region. This community was identified as one of the four highest priority communities because of its poor conditions. In order to receive the IDA-FAO organizational support the community was obliged to form an organization, the La Guaria Association (Asociacion La Guaria), of which many community members are now a part of.

Change has come to Junquillal, for both members of their own organization (Asociacion de Desarrollo Comunal de Junquillal) and the Asociacion La Guaria. This began with support for the Junquillal community in order to help them organize themselves to stop and prevent forest fires. The community actually receives the same quantity and quality of support from the different governmental and private institutions involved as do other communities in Costa Rica. However, a two-fold multiplier effect around the new activities can be identified: firstly their participation in the joint implementation program of the country; and secondly the new institutional understanding which is starting to develop in this region. A more detail description of the process follows below.

THE FOREST AS THE NEW ENGINE OF DEVELOPMENT

The Junquillal inhabitants received a course in fire prevention in forested areas. They were initially interested in preventing or eliminating the threat to their community rather than caring very much about the forest itself. However, the training explained why the forest was important to them and how to take advantage of the different products from the forest. Initially they saw the forest only as an obstacle to agricultural development activities. With this new approach the participants passed from an institutionalized perception of the forest as a source of wood and fire-wood, to one where many products and services were recognized (see Box 1), giving them a new rationale for forest conservation. A new vision of forest was thus introduced.

Several adaptations from traditional knowledge were incorporated into the new activities they began to develop. Some of these opportu-

---

BOX 1. FOREST PRODUCTS AND SERVICES

Timber: lodging and production of timber for housing.

Wood products: wood for pulp and paper, wood for energy, firewood, charcoal, posts for fences, wood for crafts and Christmas trees.

Non-wood products from forest: medicinal herbs, dyes, ornamental plants, resins, seeds, constructions materials, jeans, chemical substances, linens, fragrances, meat and animal skins.

Conservation: the retention, creation, maintenance, reproduction and survival of animal and vegetable species.

Education: the woodland environment, biodiversity and landscape in general may serve as living laboratories and outdoor classrooms. Or we may coin the term "bio-education" which covers educational activities from kindergarten to Ph.D. research.

Free Leisure: refers to the pleasant, tranquil, desired and needed rest, vacationing or sporting activities around the woods, especially for the local population.

Eco-tourism: refers to paid leisure services in National Parks, private or public reservation areas or vacation resorts.

Maintenance of the hydrologic cycle: refers to water recharge and the maintenance of rivers. Water for human, industrial and agricultural consumption, springs, and water for scenery are dependent on forests, as is flood prevention, water transportation, and hydroelectric plants.

Soil and water quality conservation: run-off and wind erosion as well as sedimentation– which are reduced by forests–may affect the quality of soil and water.

Microclimate regulation: local and horizontal precipitation and local humidity.

Wind and noise control: forests serves as windbreaks (agriculture activities) and noise barriers (housing and vacation homes).

Carbon sink: carbon sink and fixation, protecting the global environment from climate change.

Hunting: Forests are sources of wildlife which also serve as food for rural communities as well as sport for urban vacationers.

Maintenance of biological diversity in the forest ecosystems: ecosystem resilience, maintenance of the forestry capability for reducing impacts on protected areas (buffer zones), natural history, research bank (or library) for future development (of agriculture and pharmaceutical discoveries, for instance).

Cultural and religious services: Rural and indigenous communities also have beliefs, sacred places and cultural values which should be respected. Existence value.

nities to use the forest arise as very "new" possibilities for them, and in some cases it was very difficult to change the institutional understanding and learning to this new rationale of considering the forest as a different multi-product source. Some activities started almost immediately, for instance it was considered possible to use a lot of partially burned wood for cooking, instead of looking for firewood every day, or to make use of this wood as timber. Other possibilities take more time, for instance the idea of using the forest as bank for carbon dioxide (CO_2) absorption and fixation of carbon and to sell this service internationally. But because this secondary forest was state owned (IDA property) it was necessary to obtain permission from them. After discussions about how to give such permission, some ideas developed along these lines and IDA agreed to rent the property with the 200 hectares of secondary forest to the Asociacion La Guaria, for 99 years in exchange for a nominal rent of a few hundred dollars, under the understanding that the community would protect and manage the whole forested area.

FOREST MANAGEMENT AND INCENTIVES

The community of Junquillal are changing their patterns of production. They began to change their ideas of deforesting these areas and start with the only activity they knew how to work in (pasture and agriculture), and switched to work with forest as much as possible without abandoning some agriculture activities for subsistence. Personnel from IDA itself helped people from La Florida (the old squatters) to receive some training with regard to how to manage their forest. Some technicians from the Ministry of Environment (MINAE) and IDA local offices trained them to extract wood from the forest without damaging the rest of the forest and how to move within the forest without getting lost. They also received training with regard to the construction of alleys and rampart works for the prevention of forest fires, so that if a given area of forest catches fire it would not spread so easily to other areas.

Forest engineers also gave them the idea to apply for the incentive called Forest Protection Certificates (CPB in Spanish). CPB are provided by the Government of Costa Rica to people who decide to manage their forest without harvesting timber. Instead they receive approximately 40 dollars per hectare per year for 5 consecutive years.

GRAPH 1. Cluster of Forest Junquillal de Santa Cruz, Guanacaste, Costa Rica.

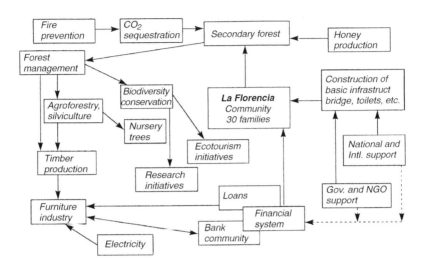

This incentive is part of the "forest services payment" approved in Forestry Law #75/5 of April 16, 1996, with funds coming from the first joint implementation transaction between Costa Rica and Norway. In future it is expected that it would be possible to collect resources from the selling of other services from forest. This is the rationale behind a law indicating "forest services" (for instance water cycle maintenance, biodiversity conservation, etc.) instead of a particular forest service (only one of the services). In other words, following approval of participation in this program, these people are receiving approximately 8 thousand dollars per year in order to provide forest services to humankind. This activity and the money it provides is creating a chain of production around the forest which did not exist before. This chain of production can perhaps best be illustrated by the cluster of activities in Junquillal illustrated in Graph 1. The payment from CPB is being used to buy building material for maintaining the fences around the forest and to pay salaries for fence maintenance and trench building against forest fires. But it also helps the community to pay a small amount as their contribution for the electrification of La Florida, Junquillal. The availability of electricity has presented the community with new opportunities as they are now able to have machinery for working with wood, pumping water and other development activities.

WOOD EXTRACTION, THE FURNITURE WORKSHOP AND OTHER ACTIVITIES

According to their own management plans only the extraction of partially burned wood from previous fires is permitted, since this forest has not been allowed to be harvested, nor has it caught fire during approximately the last three years. The extraction is done with oxen in order to reduce the impact to the rest of the forest to a minimum.

Because there is enough wood for several years, they decided to ask for assistance from the Institute of National Learning (INA, Instituto Nacional de Aprendizaje) in order to learn about furniture construction. INA trained all the inhabitants of Junquillal who wanted to participate (men and women) in handicraft production, wood carving, and the construction of windows and doors. As part of the training a small workshop was constructed. FAO provided a loan/donation for the necessary equipment for them to start their work and the money will be repaid to a revolving fund in their own bank–Bancomunal (described below).

FAO also gave a cow to each one of the new families in order to provide them with milk for their children, on the understanding that this would be repaid through small installments over several years to Bancomunal.

Additionally, as part of the forest management activities they also received training from INA with regard to apicultural activities. They received thirty beehives with which to initiate activities. The women have been trained in bottling the honey for menfolk to sell in grocery stores in the nearby town.

ELECTRICITY IN TOWN

In order to work with the machinery in the furniture workshop it was necessary to have electricity, therefore, the whole community decided to use part of the money from CPB to finance a down payment of an electricity network. Installation and the electricity service was contracted to an electrification cooperative in Guanacaste, COOPE-GUANACASTE (Cooperativa de electrificacion de Guanacaste). Additionally each family that participated contributed 15 thousand colones (approximately 60 dollars). Their residential connection and their own meter was also sold by COOPEGUANACASTE and is being repaid each month as part of the electricity bill. Unfortunately, some families are far away from the center of the town and from the electricity lines; these families still do not have electricity nor enough money to pay for the service connection.

RESEARCH AND REFORESTATION

Most of the people who received a plot of land from IDA are now dedicating some small areas to reforestation and tree nurseries. They decided that because they were not going to harvest the secondary forest, and that in the future they or future generations are going to need wood, this would be a good solution. They have four hectares of forest where they are learning silviculture techniques and are planting and testing the adaptability of some native trees from this zone, such as ron-ron (*Astronium graveolens*), pochote (*Pochota quinata*) y teca (*Tectona grandis*). They are also experimenting with some agroforestry techniques on their own properties and hope to develop more knowledge with regard to this.

Additionally some of the people from the Asociacion La Guaria are trying to enrich patches of the forest. They are planting neem trees, which contain an active ingredient that can be used as a pesticide.

BIODIVERSITY AND ECO-TOURISM

The reduction in deforestation and the adoption of fire prevention activities has allowed the secondary forest to grow more naturally with less disturbance. As a result, wildlife is also returning which includes squirrels, deer, iguanas and various species of birds.

Some people, students from universities and technical schools among others, are interested in visiting this area, while others want to know more about the social, ecological, and economic experience this community is developing. The Association is considering the possibility to apply for permission to build a lodge for visitors. This permission must be obtained from the Tourism Institute and must comply with several conditions in order to receive a tourist and be approved as tourist place. The project is currently under consideration by the Association and if approved it will go to the National Institute for approval.

BANCOMUNAL

The Junquillal inhabitants are also creating a bank called Bancomunal. This is really an endowment fund which is managed by the members of the community who contribute deposits and participate in record keeping, administrative and accounting work. The group from La Florida have a kind of Bancomunal where they deposited the money from the CPB and they are also depositing the FAO loan/donations. A group of four community members and one from the bank administers the money, approves loans for the furniture workshop and to families, and keeps records of payments and non-payments. The deposits also earn interest which allows the capital value to grow.

OTHER ACTIVITIES

There are other organizations participating in several activities which are not directly related to the forest resources, but are indirectly related because Junquillal's new dynamism. For instance the non-governmental organization (NGO) called World Vision (Vision Mundial) give the community the necessary material for building a bridge over the nearby river. At present no bridge exists and this constituted a danger for people when crossing it, especially in the rainy season. The Ministry of Road Construction and Transport (Ministerio de Obras Publicas y Transportes) helped them with the design and inspection, and the Social Assistance Institute (IMAS, Instituto Mixto de Ayuda Social) is providing the necessary salaries (160 dollars approximately per month) to the workers from the community who are building the bridge. This NGO is also helping them to build a wheel for potable water and to buy a water pump and the necessary pipes to make connections and bring water to several houses in the community.

An NGO called America's Friends (Amigos de las Americas) in collaboration with the Ministry of Health and an organized group of twelve young boys and girls from the Junquillal community are building 60 toilets, 20 dry ones and 40 wet ones. The beneficiaries dig two meter holes where they are going to install the toilet. The Ministry of Health with some external donations provides the materials and one person to supervise and advise. This youth group together with Amigos de las Americas (a small group of youths participating in grassroots activities) carry out the rest of the work.

PROBLEMS AND LIMITATIONS

All these successful initiatives are also facing a large number of problems and limitations. One of the most important bottlenecks is the absence of the potential to commercialize the craft and furniture products. La Florencia, Junquillal is far from the nearest town and the road is very bad. Regular transport for merchandise does not exist and must be previously contracted, which is therefore very expensive. This barrier to the commercialization creates production disincentives as well as discouraging new activity in general. This also results in outward migration in search of better conditions in the Central Valley and Limon, especially by those who have recently acquired new skills. Of the twenty people who finished the furniture course only 2-3 continue producing windows and doors. Trained women abandoned these activities, in spite of the fact that they were producing the most interesting and beautiful carvings.

Members of the community complain because of deforestation in the nearby forest. Though the secondary forest they are protecting has been neither burnt nor deforested by them, there are other people legally and illegally cutting trees. They argue that while they are taking care of their forest, the Santa Cruz Municipality grants forest harvesting licenses with little consideration of forest management criteria and in some cases inspectors have been bribed. Additionally the Municipality and the MINAE do not have the enforcement capabilities to control deforestation. In this sense the community has sacrificed what they consider to be an option for survival–agricultural activities on forest land and the selling of timber–while others continue to deforest.

Other people who were supposed to repay the cow loan to Bancomunal have not done so and in some cases they have sold the animal. Those who have already made the repayment complain and will not support any request for further loans to the former. Additionally, due to the lack of commercialization of their products and the loss of interest in production, only some people are using the equipment. Therefore, others consider that they are using tools and machinery which partially belongs to them. In this sense some conflicts are arising within the community and their Associations. However, many community members expect that most of these problems can eventually be resolved.

LESSONS LEARNED

Joint implementation projects should be viewed from many angles and considered with respect to the issues of cost-effectiveness, environmental effects, equity, linkage dynamics and the learning effects of the specific projects. The Junquillal community, as described above, is an especially interesting example for exploring such questions and effects. Forestry projects are in this regard an especially complicated issue, in that the forest provides a whole range of services and products and, accordingly involves a wide range of actors and stakeholders. A joint implementation program in itself is not going to solve the problems at the community level. However, if accompanied by public and private initiatives, it definitely generate an important change.

A systemic and dynamic understanding of the forest system is therefore necessary in order to avoid a strict conservationist bias to the carbon sequestration projects of joint implementation. In a certain sense, we are talking about a new rationale for the forest sector, which not only includes traditional wood products, but also many services. The production and commercialization of these services also needs to be set in the context of a whole set of inter-linkages and a cluster of activities, which if understood and encouraged correctly can become an engine of development for the community and the country. Thus, in spite of our example much more research needs to be done to clarify the necessary conditions for this kind of projects to become successful at the national and the community level.

Joint implementation programs, such as in the case of Costa Rica, also allow the generation of several other activities from the same forest without affecting carbon storage. Examples include: ecotourism; the extraction of minor forest products, such as fruits, wildlife, nuts, etc.; and the use and research of biodiversity. Therefore, it seems it is possible to realize joint implementation with positive local, social and economic impacts. Additionally, each one of these activities may generate multiple income streams, because in order to sell internationally the service of carbon sequestration, there is the need for the services of, for example cartographers, Geographical Information Systems analysts, insurance companies, foresters, engineers, economists, financial system, and other specialists. An entirely new economic cluster of activities is therefore being created around the emerging new commodity of carbon services which is only just starting to be traded internationally.

In short, the activity of carbon sequestration seems to be an especially interesting alternative for less developed countries, though it is also attractive for developed ones, because it not only creates jobs and increases income but also helps to reduce deforestation and/or may increase reforestation activities.

REFERENCES

CIFOR (1996): CIFOR's Strategy for Collaborative Forestry Research. Center for International Forestry Research, Bogor, Indonesia.

FAO (1996): FAO's First State of the World's Plant Genetic Resources: Erosion of Biodiversity and Loss of Genes Continues; Many Genebanks Threatened. Press 96/9, Rome.

Funtowicz, S. O. & Ravetz, J. R. (1991): A New Scientific Methodology for Global Environmental Issues. In: R. Costanza (Ed.): Ecological Economics. The Science and Management of Sustainability. New York: Columbia University Press. IEA (1994): World Energy Outlook. Paris.

Killingland, T. (1994): Den Nord-Amerikanske miljobevegelses syn pa Joint Implementation som virkemiddel for a redusere utslipp av klimagasser. Policy Note 1994:3, CICERO, Oslo University.

Lindegaard, Klaus and Segura, Olman. (1997): Trade Offs in Joint Implementation Strategies: The Central American Forestry Case. Paper presented at the Cross-Cultural Protection of Nature and the Environment Humanities Research Center: Man and Nature, Hollufgaard, Denmark. May 5-7, 1997.

Lovejoy, T. E. (1997): Lessons from a small country. The Washington Post, April 22, p. A-19

OECD (1995): Approaches to Dispute Settlement in Environmental Conventions and Other Legal Instruments. OECD Working Papers No. 95, Paris.

Ostrom, E. (1990): Governing the Commons: The Evolution of Institutions for Collective Action. Cambridge: Cambridge University Press.

Segura, O. et al. (1996): Politicas Forestales en Centro America: Restricciones para el desarollo del sector. CCAB-AP, San Jose, Costa Rica.

Selrod, R., Ringius, L., Torvanger, A. (1995): Joint Implementation–A Promising Mechanism for All Countries? Policy Note 1995:1, CICERO, Oslo University.

Tico Times (1997): C.R. Sells First Carbon Bonds to Norway. February 14, p. 10.

Torvanger, A. (1993): Prerequisites for Joint Implementation Projects Under the UN Framework Convention on Climate Change. Policy Note 1993.3, CICERO, Oslo University.

United Nations Development Program (UNDP) (1996): Human Development Report 1996. New York: Oxford University Press.

United Nations (1997): Kyoto Protocol to the United Nations Framework Convention on Climate Change. FCCC/CP/1997/L.7/Add.1. 10 December.

Zylicz, T. (1991): The Role of Economic Incentives in International Allocation of Abatement Effort. In: R. Costanza (Ed.): Ecological Economics. The Science and Management of Sustainability. New York: Columbia University Press.

PART II:
AMAZONIA FORESTS

Effects of Land Use and Forest Management on the Carbon Cycle in the Brazilian Amazon

Philip M. Fearnside

SUMMARY. Deforestation in the Brazilian Legal Amazon releases substantial amounts of greenhouse gases. Net committed emissions (the long-term result of emissions and uptakes in a given area that is

Philip M. Fearnside is affiliated with the National Institute for Research, Amazon-INPA, C.P. 478, 69011-970 Manaus-Amazonas, Brazil (E-mail: pmfearn@inpa.gov.br).

The author would like to thank the National Council of Scientific and Technological Development (CNPq) (BPP 350230/97-98) and the National Institute for Research in the Amazon (INPA) (PPI 5-3150) for financial support.

This paper was presented at "Conferência Internacional–Dimensões Humanas da Mudança Climática Global e do Manejo Sustentável das Florestas das Américas: Uma Conferência Interamericana," December 1-3, 1997, Departamento de Economia, Universidade de Brasília, Brasília.

[Haworth co-indexing entry note]: "Effects of Land Use and Forest Management on the Carbon Cycle in the Brazilian Amazon." Fearnside, Philip M. Co-published simultaneously in *Journal of Sustainable Forestry* (Food Products Press, an imprint of The Haworth Press, Inc.) Vol. 12, No. 1/2, 2001, pp. 79-97; and: *Climate Change and Forest Management in the Western Hemisphere* (ed: Mohammed H. I. Dore) Food Products Press, an imprint of The Haworth Press, Inc., 2001, pp. 79-97. Single or multiple copies of this article are available for a fee from The Haworth Document Delivery Service [1-800-342-9678, 9:00 a.m. - 5:00 p.m. (EST). E-mail address: getinfo@haworthpressinc.com].

79

cleared) totaled 267-278 million t of CO_2-equivalent carbon in 1990 (under low and high trace gas scenarios), while the corresponding annual balance of net emissions (the balance in a single year over the entire region, including areas cleared in previous years) in 1990 was 354-358 million t from deforestation plus 62 t from logging. These figures contrast sharply with official pronouncements that claim little or even no net emission from Amazonia. Most emissions are caused by medium and large ranchers (despite recent official statements to the contrary), a fact which means that deforestation could be greatly slowed without preventing subsistence clearing by small farmers. The substantial monetary and non-monetary benefits that avoiding this impact would have provide a rational for making the supply of environmental services a long-term objective in reorienting development in Amazonia. *[Article copies available for a fee from The Haworth Document Delivery Service: 1-800-342-9678. E-mail address: <getinfo@haworthpressinc.com> Website: <http://www.haworthpressinc.com>]*

KEYWORDS. Deforestation, emissions, subsistence clearing, environmental services, development

INTRODUCTION

Controversies Surrounding Brazilian Emissions

Brazil's current and potential future emissions of greenhouse gases from deforestation in Amazonia are both items of global concern and controversy. The numbers that have been presented by different authorities for the magnitude of these emissions range from zero to values on a par with the total emission by the world's fleet of automobiles. In the face of such discrepancies, it is common for those not closely following the issues to either postpone accepting any value 'until the experts agree' (i.e., the observer will continue to act as if the impact were zero), or to assume that the midpoint of the various values that have been presented to the public represents the best estimate. Neither reaction is advisable: there is no substitute for taking the time to understand the issues involved and to evaluate the appropriateness and reliability of the different numbers available. One must then have the courage to act on the basis of the best estimate, once it has been identified based on its merits. The range of genuine scientific uncertainty surrounding the emissions estimates is very much less than the range of statements that have been made on the matter because many of the existing values contain known errors or omissions.

In addition to controversies about how many tons of gases are emitted, there is an equally wide range of opinion as to whether a given level of emissions represents an insignificant dribble or a major catastrophe. Unfortunately, the information in the present paper indicates that the emissions from Amazonian deforestation are large and their impact is important. How climate negotiations are handled can determine whether this major impact represents bad news for the population of the Amazonian interior, or whether it represents an opportunity to turn the environmental service of avoiding greenhouse gas emissions into a sustainable means of supporting that population.

Magnitude of Brazilian Emissions

The values obtained for the magnitude of Brazilian emissions depend on the values used for basic parameters such as deforestation rate, biomass, and carbon uptake by the replacement landscape. They also depend on the inclusion or omission of different portions of the emission, such as emission from decomposition, emissions from reburns (burns other than the initial one), emissions from underground biomass, soil carbon, hydroelectric reservoirs, and the effect of trace gases such as methane and nitrous oxide.

Some very high estimates of emissions from Brazilian Amazonia resulted from an estimate of deforestation rate at 200,000 km²/year (WRI, 1990: 103). This deforestation rate estimate was taken from an estimate of area burned (which is not the same thing as deforestation) for 1987 derived by Setzer et al. (1988) and extrapolated to the decade of the "1980s." Both technical errors in the deforestation rate estimate and the extrapolation from an atypical year (1987) invalidate this emissions estimate (Fearnside, 1990a). Another high estimate uses a deforestation rate estimate of 50,000 km²/year (Myers, 1989, 1991) based on a preliminary version of an estimate by Setzer and Pereira (1991) which estimated 48,000 km²/year as the rate for 1988. The 50,000 km²/year rate (Myers, 1989, 1991) was also used as the deforestation rate estimate in emissions calculations by Houghton (1991). This deforestation rate estimate also suffers from known technical errors that inflate the resulting value (see Fearnside, 1990a). The best current estimate for the average 1980-1989 deforestation rate is 20.3 × 10³ km² (based on Fearnside, 1997a). This and other deforestation rate values given in this paper refer to loss of "forest" (as defined in Fearnside and Ferraz, 1995), and do not include loss of the *cerrado* (central Brazilian scrubland), or degradation of forest through logging or other processes.

Biomass estimates vary greatly both in their magnitude and in the reliability of the data and of the calculation procedure. An estimate of average total biomass (including underground biomass) of only 155.1 t/ha (expressed as dry weight biomass, not carbon) was derived by Brown and Lugo (1984). This value, which is less than half as large as current estimates of this parameter, was used by Detwiler and Hall (1988) to estimate emissions from tropical deforestation. Although the biomass estimate is defended by no one, including its original authors, it is still relevant today because it forms part of the basis of the estimate by the Intergovernmental Panel on Climate Change (IPCC) of 1.6 Gt (gigatons = 10^9 t) of carbon as the global total net emission from tropical land-use change (Schimel et al., 1996: 79). The 1.6 Gt C global value for emissions from tropical deforestation in the 1980-1989 period was originally derived (Watson et al., 1990: 11) as the midpoint between a low estimate of 0.6 Gt C/year by Detwiler and Hall (1988: 43) and a high estimate of 2.5 Gt C/year by Houghton et al. (1987: 125), the latter of which used an estimate for total biomass of forest of 352 t/ha from Brown and Lugo (1982). In the 1990 IPCC report (Houghton et al., 1990) the 1.6 Gt C/year value was called the "land-use change term," but emissions from sources other than tropical deforestation were, in effect, all considered to have zero values. In the 1995 report (the Second Assessment Report, or SAR: Houghton et al., 1996), the 1.6 Gt C/year term was explicitly restricted to tropical deforestation, and a separate-0.5 Gt C/year term was added to represent carbon uptake by forest growth in the temperate zone. The 1.6 ± 1 Gt C/year term for tropical deforestation was maintained in the SAR (Schimel et al., 1996: 79) based on approximate agreement with an estimate of 1.65 ± 0.4 Gt C/year by Brown et al. (1996: 777). The latter estimate is based primarily on an estimate by Dixon et al. (1994), which used biomass estimates for Brazilian Amazonia based on Fearnside (1992b): 272 t /ha, or about 33% lower than current estimates for biomass being cleared (Fearnside, nd; updated from Fearnside, 1997b). In addition, the Dixon et al. (1994) estimate was, in the case of Brazilian Amazonia, based on a deforestation estimate for the 1980s (Skole and Tucker, 1993) that underestimates the rate of clearing in that period by 24% (Fearnside, 1993a). Clearly, these differences are sufficient to make a substantial difference in final conclusions regarding the magnitude of greenhouse gas emissions from deforestation.

Net committed emissions expresses the ultimate contribution of transforming the forested landscape into a new one, using as the basis of comparison the mosaic of land uses that would result from an equilibrium condition created by projection of current trends. This includes emissions from decay or reburning of logs that are left unburned when forest is initially felled (committed emissions), and uptake of carbon from growing secondary forests on sites abandoned after use in agriculture and ranching (committed uptake) (Fearnside, 1997b).

Net committed emissions considers the emissions and uptakes that will occur as the landscape approaches a new equilibrium condition in a given deforested area. Here the area considered is the 13.8×10^3 km^2 of Brazil's Amazonian forest that was cut in 1990, the reference year for baseline inventories under the United Nations Framework Convention on Climate Change (UN-FCCC). The "prompt emissions" (emissions entering the atmosphere in the year of clearing) are considered along with the "delayed emissions" (emissions that will enter the atmosphere in future years), as well as the corresponding uptake as replacement vegetation regrows on the deforested sites. Not included are trace gas emissions from the burning and decomposition of secondary forest and pasture biomass in the replacement landscape, although both trace gas and carbon dioxide fluxes are included for emissions originating from remains of the original forest biomass, from loss of intact forest sources and sinks, and from soil carbon pools. Net committed emissions are calculated as the difference between the carbon stocks in the forest and the equilibrium replacement landscape, with trace gas fluxes estimated based on fractions of the biomass that burn or decompose following different pathways.

In contrast to net committed emissions, the annual balance considers releases and uptakes of greenhouse gases in a given year (Fearnside, 1996). Annual balance considers the entire region (not just the part deforested in a single year), and considers the fluxes of gases entering and leaving the region both through prompt emissions in the newly deforested areas and through the "inherited" emissions and uptakes in the clearings of different ages throughout the landscape. Inherited emissions and uptakes are the fluxes occurring in the year in question that are the result of clearing activity in previous years, for example, from decomposition or reburning of the remaining biomass of the original forest. The annual balance also includes trace gases from secondary forest and pasture burning and decomposition.

The annual balance represents an instantaneous measure of the fluxes of greenhouse gases, of which carbon dioxide is one. Even though the present calculations are made on a yearly basis, they are termed "instantaneous" here to emphasize the fact that they do not include future consequences of deforestation and other actions taking place during the year in question.

FOREST BIOMASS

Emissions of greenhouse gases from deforestation are essentially proportional to the biomass of the forest. The wide range of estimates for biomass is therefore a key factor in the range of values that different authors have calculated. In a number of cases, however, underestimates of biomass have been used in conjunction with overestimates of deforestation rate. In such cases, the errors may cancel each other out, and can produce emissions estimates that fall in a reasonable range. However, agreement among estimates that differ in their underlying assumptions and parameters is illusory and misleading, as it does not indicate replication. It is important to base policies on estimates that not only have the right final result, but that reach their result for the right reasons–that is, based on the best current estimates of all parameters.

A series of estimates has been produced by Sandra Brown and Ariel Lugo (Brown and Lugo, 1982, 1984, 1992a, 1992b, 1992c; Brown et al., 1989), while I have produced a series of estimates with substantially higher values (Fearnside, 1985, 1986, 1987a, 1990b, 1991, 1992a, 1992b, 1994, 1997b, nd). It is very important to understand why the differences exist. The very low estimate of 155.1 t/ha, 133.7 t/ha of which was above-ground (Brown and Lugo, 1984) apparently contained calculation errors, since the original FAO data used in that estimate yield higher values when the published calculation procedure is applied (see Fearnside, 1987a, 1986). Brown and Lugo themselves revised the above-ground portion of their estimate upward by 27% to 169.68 t/ha in a subsequent publication (Brown et al., 1989). However, this and subsequent estimates of above-ground biomass of 162 (Brown and Lugo, 1992a) and 227 t/ha (Brown and Lugo, 1992b) contain substantial omissions (see Fearnside, 1992a, 1993b). These include a +15.6% adjustment of above-ground live biomass for form factor, +12.0% for trees < 10 cm diameter at breast height (DBH), +3.6% for trees 30-31.8 cm DBH, +2.4% for palms, +5.3% for vines, +0.2% for other non-tree components, − 0.9% for bark volume and density, and − 6.6% for hollow trees. These adjustments to above-ground live biomass total +31.7%. The total so obtained must then be increased with additions for dead biomass (8.6%) and for below-ground biomass (33.6%) (Fearnside, nd, updated from Fearnside, 1994; see Fearnside, 1997b). The current estimates (Fearnside, nd, updated from Fearnside, 1994; see Fearnside, 1997b) are based on much more data than earlier estimates, using 2954 ha of data spread throughout the Legal Amazon region in 1-ha forest inventory plots. Approximately 90% of the data are based on the RADAMBRASIL surveys, and the remaining 10% on FAO data. The current biomass estimate incorporates improved estimates of the average basic density of wood, desegregated by forest type (Fearnside, 1997c).

GREENHOUSE GAS EMISSIONS

Brazil's official estimates of greenhouse gas emissions have produced some extraordinarily low values. On the eve of the 1992 United Nations Conference on Environment and Development (UNCED), or "ECO-92," in Rio de Janeiro. Brazil's National Institute for Space Research (INPE) announced that Brazilian deforestation released only 1.4% of the world's CO_2 emissions (Borges, 1992), a value about three times lower than that derived in the current paper. Such a low value was obtained by counting only prompt emissions released through the initial burning of the forest, ignoring decomposition and re-burns. Only 39% of the gross release of above-ground carbon, or 27% of the gross release of total carbon (including below-ground biomass and soil carbon) occurs through this pathway for the carbon dioxide component of net committed emissions (Fearnside, 2000b, updated from Fearnside, 1997b).

On the eve of the 1997 conference of the parties to the UN Framework Convention on Climate Change (UN-FCCC), INPE announced that Brazil releases zero net emissions from deforestation (ISTOÉ, 1997). This extraordinary conclusion was apparently reached by ignoring all emissions other than the initial burn, combined with the belief that the crops planted can somehow absorb this amount of carbon. INPE claimed that "the crops that grow wind up absorbing the carbon that was thrown into the atmosphere by the burning" (ISTOÉ, 1997). Unfortunately, only 7% of the net committed emissions are reabsorbed by the replacement landscape (Fearnside, 1997b; see also Fearnside and Guimaràes, 1996).

Current estimates of the 1990 emission from deforestation in the Brazilian Legal Amazon are given in Table 1 in terms of net committed emissions and annual balance. Two scenarios are given: "low" and "high" trace gas emissions. These represent a range of emissions factors, or the amount of each gas emitted by different processes such as flaming and smoldering combustion. The range of doubt concerning other important processes, such as forest biomass and deforestation rate at different locations, is not included. The annual balance was higher than the net committed emissions in 1990 because deforestation rates had been higher in the years immediately preceding this year were higher, therefore leaving larger quantities of unburned biomass to produce emissions in the years that follow. My current best estimate for 1990 (Table 1) is 267×10^6 t C of net committed emissions and 354×10^6 t C of annual balance from deforestation, plus an additional

TABLE 1. Comparison of Methods of Calculating the 1990 Global Warming Impact of Deforestation in Originally Forested Areas Brazilian Amazonia in Millions of Tons of CO_2-Equivalent Carbon

Scenario	Gases included	Net committed emissions (Deforest-ation only) (a,b)	Annual balance		
			Deforestation (b) only	Logging	Deforstation (b) + logging
Low trace gas	CO_2 only	255	329	61	390
	CO_2, CH_4, N_2O (c)	267	354	62	416
High trace gas	CO_2 only	255	324	61	385
	CO_2, CH_4, N_2O (c)	278	358	62	421

(a) Intinite time horizon for fluxes from biomass, soil C and replacement vegetation uptake 100-year time horizon for recurrent fluxes (cattle, pasture soil N_2O, hydroelectric CH_4 and losses of intact forest sources and sinks); 100-year non-coterminous time horizons for impacts; no discounting.

(b) For clearing in originally forested areas only (does not include *cerrado* clearing).

(c) CO, NO_x and NMHC are also included in the analysis, but the IPCC SAR global warming potentials for these gases are equal to zero.

61×10^6 t C from logging (Fearnside, nd). Trace gases are accounted for using the 100-year integration global warming potentials adopted by the IPCC's second assessment report (Schimel et al., 1996). Only deforestation (that is loss of forest, including both clearing and flooding by hydroelectric dams) is given here, not loss of *cerrado* (the central Brazilian scrubland that was the original vegetation in about 20% of the Legal Amazon).

The relative weight of small farmers versus large landholders in Amazonian deforestation is continually subject to change as a result of changing economic and demographic pressures. The behavior of landholders is most sensitive to economic changes such as the interest rates offered by money market and other financial investments, government subsidies for agricultural credit, the rate of general inflation, and changes in the price of land. Tax incentives were a strong motive in the 1970s and 1980s. In June 1991 a decree (No. 153) suspended the granting of *new* incentives. However, the old (i.e., already approved) incentives continue to the present day, contrary to the popular impression that was fostered by numerous statements by government officials to the effect that incentives had been ended. Most of the other forms of incentives, such as government-subsidized credit at rates far below those of Brazilian inflation, effectively dried up after 1984 (the last year, for example, when the SUFRAMA ranches north of Manaus made significant clearings).

For decades preceding the initiation of the Plano Real in 1994, hyperinflation was the dominant feature of the Brazilian economy. Land played a role as store of value, and its value was bid up to levels much higher than what could be justified as an input to agricultural and ranching production. Nevertheless, vast fortunes were made in Amazonian land, and deforestation played a critical role as a means of holding claim to speculative investments in land (see Fearnside, 1987b).

The Plano Real sharply cut the rate of inflation in Brazil. Brazil Fundação Getúlio Vargas has found that land values reached a peak in 1995, and subsequently fell substantially in 1996 and 1997 (*O Diário,* 25 January 1998). This is a likely explanation of a decline in deforestation rate over the 1995-1997 period indicated by LANDSAT data recently released by INPE. These data indicate a peak of annual deforestation in 1995 of 29.1×10^3 km^2, followed by 18.2×10^3 km^2 in 1996 and a preliminary estimate of 13.0×10^3 km^2 in 1997 (Brazil,

INPE, 1998). The peak in 1995, which is a jump from the already very high rate of 14.9 × 10^3 km^2 in 1994, is probably in large part a reflection of economic recovery under the Plano Real, and consequently the availability of larger volumes of money to be invested in cattle ranches.

INTERPRETING VALUES FOR EMISSIONS IMPACT

Apportioning the Blame Among Agents

An important feature of the problem of greenhouse gas emissions from deforestation is that forest clearing could be greatly curtailed without provoking tremendous social impacts. This is because most of the clearing is done by large or medium ranches rather than by small farmers: only 30.5% of the clearing in 1990 and 1991 is attributable to small farmers (Fearnside, 1993c). The idea that rainforests are being cleared by poor shifting cultivators who would go hungry if forced to stop is largely inappropriate for Brazilian Amazonia, where almost 70% of the clearing is done by the rich. In addition, national agricultural production is not heavily dependent on clearing more Amazonian forest because most of the cleared area becomes low-quality pasture that degrades after only about a decade. Only 6% of the value of Brazil's agricultural production comes from Amazonia, and the vast majority of the 470,000 km^2 (an area the size of France) already deforested by 1994 is either cattle pasture or secondary forest in abandoned pasturelands. Lack of space in the already-deforested portion of the region does not limit implanting higher yielding systems of both commercial agriculture and of food crops for feeding subsistence farmers.

The proportion of the region's deforestation done by landholders of different sizes (based on Fearnside, 1993c) can be used to attribute responsibility for greenhouse gas emissions among different classes of actors. Contrary to recent statements by the head of the Brazilian Institute for Environment and Renewable Natural Resources (IBA-MA) (Traumann, 1998), deforestation data for 1995 and 1996 released by INPE (Brazil, INPE, 1998) do not indicate that small farmers are now the primary agents of deforestation. The fact that about half (59% in 1995 and 53% in 1996) of the area of new *clearings* (as distinct from the area of the *properties* in which the clearings were located) have areas under 100 ha reinforces the conclusion that most of the deforestation is being done by large ranchers, as no small farmer can clear anywhere near 100 ha in a single year. Only 21% of the area of new clearings in 1995 and 18% in 1996 were under 15 ha. Small farmer families are only capable of clearing about 3 ha/year with family labor (Fearnside, 1980), and this is reflected in deforestation behavior in settlement areas (Fearnside, 1984).

Table 2 shows that one large rancher (with 1000 ha or more of land) has a greater impact on global warming than 273 small farmers (with < 100 ha of land), or over 3800 people in Brazil's cities. This dramatizes the tremendous environmental impact wreaked by a minuscule fraction of Brazil's population. This fact provides the key to taking measures to slow deforestation without provoking unacceptable social impacts, and turning environmental services such as avoiding global warming into a means of supporting the rural population of the region (Fearnside, 1997d). In what I term the "Robin Hood" solution, the value of the environmental change now being caused by the rich could be used to support the poor. A long list of hurdles would have to be crossed to turn environmental services into a form of sustainable development for rural Amazonia (Fearnside, 1997d). Nevertheless, priority must be given to creating the scientific, institutional and diplomatic basis for this if we are ever to attain the long-term objective of using environmental services as the basis of support for the population instead of the current systems based on traditional commodities such as timber and beef.

Avoided Emissions versus Stock Maintenance

Environmental services include maintenance of biodiversity and water cycling, as well as the benefits for mitigating global warming that are the subject of this paper. The value attributed to global warming benefits depends greatly on the way in which the credits are calculated. Negotiations under the UN-FCCC so far recognize only incremental changes in flows of carbon; in other words, credit for "avoided emissions" can only be gained from avoiding deforestation if a given tract of forest would have been cut down in the absence of a mitigation program (see Fearnside, 1995). This is also the criterion applied by the Global Environment Facility in assessing the carbon benefits of projects financed with the objective of combating global warming.

TABLE 2. Greenhouse Impact per Capita

Source	Population (million)	Low trace gas scenario			High trace gas scenario		
		Annual emission (million t CO_2 equivalent C) (b)	Annual emission per capita (t CO_2 equivalent C)	Number of people needed to equal one large rancher	Annual emission (million t CO_2 equivalent C) (b)	Annual emission per capita (t CO_2 equivalent C) (b)	Number of people needed to equal one large rancher
Brazil:							
Large rancher population of Amazonia (a)	0.1	95	693.0	1	189	1,382.4	1
Medium-sized rancher population of Amazonia (a)	0.5	105	219.1	3	81	167.8	8
Small farmer population of Amazonia (a)	6.7	88	13.2	53	34	5.1	273
Rural Amazonia total	8	287	43.2	16	303	37.9	37
Rest of Brazil	132	47	0.4	1,946	47	0.4	3,882
Brazil total	140	655	4.7	148	680	4.9	285
World	5,300	7,996	1.5	459	8,074	1.5	907
United States	210	1,060	5.0	137	1,060	5.0	274

(a) "Large ranches" are > 1,000 ha in area, "middle-sized ranches" are 100-1,000 ha in area, "small farms" are < 100 ha in area. The 1990 rural population is apportioned between these categories in proportion to the number of establishments censused in 1985.

(b) Emissions are allocated among property classes in accord with their proportion of the 1990 clearing activity by each class in the Legal Amazon as a whole.

Policies that result in maintenance of Amazonian forest provide two types of service in averting global warming: one is immediate reduction of the fluxes of greenhouse gases to the atmosphere, the other is avoidance of the much larger cumulative impact that would occur were Brazil's vast remaining tracts of forest felled in the future. The current methodology based on "net incremental costs" refers only to the first of these benefits. Maintenance of carbon stock receives no credit. However, strong arguments exist for rewarding this service, since the consequences of not maintaining the forest would be severe. Deforestation is a process that tends to become more difficult to stop once it gets underway in an area. While many tropical forests around the world have already been reduced to small remnants, Brazil was estimated by the FAO forest resources assessment to contain 41% of all tropical "rainforest" remaining in the world in 1990 (FAO, 1994).

An objection frequently raised with respect to recognizing maintenance of carbon stocks by tropical forests as a service, as opposed to reduction of carbon flows, is that countries with large fossil fuel deposits would then demand compensation for the unexploited stocks they hold. However, there are two fundamental differences between carbon stocks in tropical forests and those in fossil fuels. One is that most of the approximately 5000 Gt of carbon in fossil fuel deposits (Perry and Landsberg, 1977 cited by Bolin et al., 1979: 33) is not really 'at risk,' since most of it is not likely to be burned in the foreseeable future (the world currently burns approximately 6 Gt of fossil fuel carbon annually). Tropical forests, on the other hand, could quite easily be completely cleared within a century. The other difference is that fossil fuel use can be relatively easily controlled through economic instruments such as taxes and tariffs; it is not necessary to place guards at the oil wells to keep people from pumping the oil. Tropical forests, on the other hand, require more active measures if they are to be kept standing. Attributing value to the service of maintaining carbon stocks in tropical forest is fundamental to creating the motivation to take the necessary steps to assure that they are not cut. It is also worth noting that maintaining carbon in tropical forests has other benefits in maintaining biodiversity at the same time, while maintaining carbon stocks in fossil fuel deposits does not. The carbon benefits of maintaining these stocks are completely reversible (an atom of carbon is the same, regardless of its source, and it can be removed from the atmosphere by incorporation into biomass elsewhere). Biodiversity, however, is not interchangeable, and once eco-systems are destroyed and/or species are driven extinct, they are not recovered.

Discounting Carbon

Discounting of carbon benefits is another feature of accounting for the benefits that can significantly affect the conclusions. At present, the GEF does not discount any physical parameters, such as carbon, in assessing the benefits of proposed mitigation projects: a ton of emission avoided today has the same benefit as a ton avoided 20 years from now. However, good reasons exist for giving some credit for carbon benefits on the short term as opposed to the long term. Global warming is not a one-time environmental catastrophe. Rather, with each degree of warming the probability increases that given levels of impacts will occur from that time onwards. If a given amount of warming is postponed from an earlier year to a later year, then all of the increased impacts (including human deaths) that would have occurred between the earlier year and the later year represent a real gain. This gain should be viewed as a *permanent* gain, even though the same impacts could be expected to happen anyway shortly after the delay period. The logic is the same as that used crediting greenhouse gas emission avoidance by reducing the use of fossil fuels: reducing the consumption of a oil by one barrel in a given year is considered a permanent savings, even though the same barrel of oil may be pumped out of the ground and burned the following year. This is because the burning of all subsequent barrels of oil is also delayed by one year.

Discounting benefits gives more weight to carbon emissions from deforestation as compared to those from fossil fuels. This is because fossil fuel emissions are almost all in the form of CO_2, which has a modest radiative forcing (an instantaneous measure of the amount of heat that the gas prevents from being re-radiated to outer space), but each molecule remains in the atmosphere for approximately 120 years (Shine et al., 1990: 60). Deforestation emits most of its carbon as CO_2, but, unlike fossil fuel combustion, some of the carbon is emitted as CH_4, which has a greater impact per ton of carbon but which remains in the atmosphere for an average of only 12.2 years (Schimel et al., 1996: 121). In addition, the IPCC does not currently count indirect effects of CO (a gas which lengthens the lifetime of CH_4 in the atmosphere by the removing OH radicals that degrade methane). Inclusion of these effects in future revisions of the accounting procedures would further increase the effect of discounting on deforestation impacts as compared to fossil fuel impacts. Forest loss through flooding by hydroelectric dams has substantially greater impact relative to thermo-electric energy production if discounting is applied (Fearnside, 1997e).

The IPCC currently expresses the relative impact of different greenhouse gases by global warming potentials (GWPs), which express the impact of a single pulse of each gas relative to a simultaneous pulse of an equal weight of CO_2 (Schimel et al., 1996). Time horizons are considered of 20, 100 and 500 years, without discounting. Most emphasis in the policy discussions is given to the 100-yr time horizon, especially in the executive summaries. The 20- and 500-year time horizons make the middle value of 100 years appear reasonable through a sort of "Goldilocks effect," but, in reality, there is little justification for attributing equal weight to effects over the course of periods as long as 100 years (let alone 500 years). Changes occurring in year one have more importance than those occurring in year 99 not only as a result of a selfish perspective on the part of the current generation, but also because of the benefits of delaying the stream of impacts provoked by temperature increase, as mentioned earlier.

While many questions of policy (in addition to science) need to be resolved in selecting the way that the value of the impact of global warming is calculated, and consequently the benefit of avoiding it, the emissions from Amazonian deforestation are sufficiently large that all likely methods would lead to the conclusion that deforestation causes a significant global impact. Avoiding global warming, together with other environmental services in maintaining biodiversity and the regional hydrological cycle, provide a potential basis for sustaining both the rural population of the region and the ecological functions of the tropical rain forest (Fearnside, 1997d).

CONCLUSIONS

Deforestation in the Brazilian Legal Amazon releases substantial amounts of greenhouse gases. Net committed emissions (the long-term result of emissions and uptakes in a given area that is cleared) totaled 267-278 million t of CO_2-equivalent carbon in 1990 (under low and high trace gas scenarios), while the corresponding annual balance of net emissions (the balance in a single year over the entire region, including areas cleared in previous years) in 1990 was 354-358 million t from deforestation plus 62 t from logging. These figures contrast sharply with official pronouncements that claim little or even no net emission from Amazonia. Most emissions are caused by medium and large ranchers (despite recent official statements to the contrary), a fact which means that deforestation could be greatly slowed without preventing subsistence clearing by small farmers. The substantial monetary and non-monetary benefits that avoiding this impact would have provide a rational for making the supply of environmental services a long-term objective in reorienting development in Amazonia.

REFERENCES

Bolin, B., E.T. Degens, P. Duvigneaud and S. Kempe. 1979. The global biogeochemical carbon cycle. pp. 1-56 In: B. Bolin, E.T. Degens, S. Kempe and P. Ketner (eds.) *The Global Carbon Cycle. Scientific Committee on Problems of the Environment (SCOPE) Report No. 13*. John Wiley & Sons, New York, U.S.A. 491 p.

Borges, L. 1992. "Desmatamento emite só 1,4% de carbono, diz Inpe" *O Estado de São Paulo* 10 de abril de 1992, p. 13.

Brazil, INPE. 1998. Amazônia: Desflorestamento 1995-1997. Instituto Nacional de Pesquisas Espaciais (INPE), São José dos Campos, São Paulo. Document released via internet (http://www.inpe.br).

Brown, S. and A.E. Lugo. 1982. The storage and production of organic matter in tropical forests and their role in the global carbon cycle. *Biotropica* 14(2): 161-187.

Brown, S. and A.E. Lugo. 1984. Biomass of tropical forests: A new estimate based on forest volumes. *Science* 223: 1290-1293.

Brown, S. and A.E. Lugo. 1992a. Biomass estimates for Brazil's Amazonian moist forests. pp. 46-52 In: *Forest '90: Anais do Primeiro Simpósio Internacional de Estudos Ambientais em Florestas Tropicais Umidas*. Biosfera–Sociedade Brasileira para a Valorização do Meio Ambiente, Rio de Janeiro, Brazil. 508 p.

Brown, S. and A.E. Lugo. 1992b. Aboveground biomass estimates for tropical moist forests of the Brazilian Amazon. *Interciencia* 17(1): 8-18.

Brown, S. and A.E. Lugo. 1992c. Biomass of Brazilian Amazonian forests: The need for good science. *Interciencia* 17(4): 201-203.

Brown, S., A.J.R. Gillespie and A.E. Lugo. 1989. Biomass estimation methods for tropical forests with applications to forest inventory data. *Forest Science* 35: 881-902.

Brown, S., J. Sathaye, M. Cannell, P. Kauppi, P. Burschel, A. Grainger, J. Heuveldop, R. Leemans, P.M. Moura Costa, M. Pinard, S. Nilsson, W. Schopfhauser, R. Sedjo, N. Singh, M. Trexler, J. van Minnen and S. Weyers. 1996. Management of forests for mitigation of greenhouse gas emissions. pp. 773-797 In: R.T. Watson, M.C. Zinyowera, R.H. Moss and D.J. Dokken (eds.) *Climate Chante 1995: Impacts, Adaptations and Mitigation of Climate Change: Scientific-Technical Analysis.* Cambridge University Press, Cambridge, U.K. 878 p.

Detwiler, R.P. and C.A.S. Hall. 1988. Tropical forests and the global carbon cycle. *Science* 239: 42-47.

O Diário [Mogi das Cruzes]. 25 January 1998. "Preços das terras estão caindo, afirma FGV." p. 5.

Dixon, R.K., S. Brown, R.A. Houghton, A.M. Solomon, M.C. Trexler and J. Wisniewski. 1994. Carbon pools and flux of global forest ecosystems. *Science* 263: 185-190.

FAO (Food and Agriculture Organization of the United Nations). 1993. *Forest Resources Assessment 1990: Tropical Countries.* (FAO Forestry Paper 112). FAO, Rome, Italy. 61 pp. + annexes.

Fearnside, P.M. 1980. Land use allocation of the Transamazon Highway colonists of Brazil and its relation to human carrying capacity. pp. 114-138 In: F. Barbira-Scazzocchio (ed.). *Land, People and Planning in Contemporary Amazonia.* University of Cambridge Center of Latin American Studies Occasional Paper No. 3, Cambridge, U.K. 313 p.

Fearnside, P.M. 1984. Land clearing behavior in small farmer settlement schemes in the Brazilian Amazon and its relation to human carrying capacity. pp. 255-271 In: A.C. Chadwick and S.L. Sutton (eds.). *Tropical Rain Forest: The Leeds Symposium.* Leeds Philosophical and Literary Society, Leeds, U.K. 335 p.

Fearnside, P.M. 1985. Brazil's Amazon forest and the global carbon problem. *Interciencia* 10(4): 179-186.

Fearnside, P.M. 1986. Brazil's Amazon forest and the global carbon problem: Reply to Lugo and Brown. *Interciencia* 11(2): 58-64.

Fearnside, P.M. 1987a. Summary of progress in quantifying the potential contribution of Amazonian deforestation to the global carbon problem. pp. 75-82 In: D. Athié, T.E. Lovejoy and P. de M. Oyens (eds.) *Proceedings of the Workshop on Biogeochemistry of Tropical Rain Forests: Problems for Research.* Universidade de São Paulo, Centro de Energia Nuclear na Agricultura (CENA), Piracicaba, São Paulo, Brazil. 85 p.

Fearnside, P.M. 1987b. Causes of Deforestation in the Brazilian Amazon. pp. 37-61 In: R.E. Dickinson (ed.) *The Geophysiology of Amazonia: Vegetation and Climate Interactions.* John Wiley & Sons, New York, U.S.A. 526 p.

Fearnside, P.M. 1990a. The rate and extent of deforestation in Brazilian Amazonia. *Environmental Conservation* 17(3): 213-226.

Fearnside, P.M. 1990b. Contribution to the greenhouse effect from deforestation in Brazilian Amazonia. pp. 465-488 In: *Intergovernmental Panel on Climate Change (IPCC), Response Strategies Working Group (RSWG), Subgroup on Agriculture, Forestry and other Human Activities (AFOS). Proceedings of the Conference on Tropical Forestry Response Options to Global Climate Change.* U.S. Environmental Protection Agency, Office of Policy Assessment (USEPA-OPA, PM221), Washington, D.C., U.S.A. 531 p.

Fearnside, P.M. 1991. Greenhouse gas contributions from deforestation in Brazilian

Amazonia. pp. 92-105 In: J.S. Levine (ed.) *Global Biomass Burning: Atmospheric, Climatic, and Biospheric Implications*. MIT Press, Boston, Massachusetts, USA 640 p.

Fearnside, P.M. 1992b. Forest biomass in Brazilian Amazonia: Comments on the estimate by Brown and Lugo. *Interciencia* 17(1): 19-27.

Fearnside, P.M. 1992b. *Greenhouse Gas Emissions from Deforestation in the Brazilian Amazon*. Carbon Emissions and Sequestration in Forests: Case Studies from Developing Countries. Volume 2. LBL-32758, UC-402. Climate Change Division, Environmental Protection Agency, Washington, DC and Energy and Environment Division, Lawrence Berkeley Laboratory (LBL), University of California (UC), Berkeley, California, USA 73 p.

Fearnside, P.M. 1993a. Desmatamento na Amazônia: Quem tem razão–o INPE ou a NASA? *Ciência Hoje* 16(96): 6-8.

Fearnside, P.M. 1993b. Biomass of Brazil's Amazonian forests: Reply to Brown and Lugo revisited. *Interciencia* 18(1): 5-7.

Fearnside, P.M. 1993c. Deforestation in Brazilian Amazonia: The effect of population and land tenure. *Ambio* 22(8): 537-545.

Fearnside, P.M. 1994. Biomassa das florestas Amazônicas brasileiras. pp. 95-124 In: *Anais do Seminário Emissão × Seqüestro de CO2*. Companhia Vale do Rio Doce (CVRD), Rio de Janeiro, Brazil. 221 p.

Fearnside, P.M. 1995. Global warming response options in Brazil's forest sector: Comparison of project-level costs and benefits. *Biomass and Bioenergy* 8(5): 309-322.

Fearnside, P.M. 1996. Amazonia and global warming: Annual balance of greenhouse gas emissions from land-use change in Brazil's Amazon region. pp. 606-617 In: J. Levine (ed.) *Biomass Burning and Global Change. Volume 2: Biomass Burning in South America, Southeast Asia and Temperate and Boreal Ecosystems and the Oil Fires of Kuwait*. MIT Press, Cambridge, Massachusetts, USA 902 p.

Fearnside, P.M. 1997a. Monitoring needs to transform Amazonian forest maintenance into a global warming mitigation option. *Mitigation and Adaptation Strategies for Global Change* 2(2-3): 285-302.

Fearnside, P.M. 1997b. Greenhouse gases from deforestation in Brazilian Amazonia: Net committed emissions. *Climatic Change* 35(3): 321-360.

Fearnside, P.M. 1997c. Wood density for estimating forest biomass in Brazilian Amazonia. *Forest Ecology and Management* 90(1): 59-89.

Fearnside, P.M. 1997d. Environmental services as a strategy for sustainable development in rural Amazonia. *Ecological Economics* 20(1): 53-70.

Fearnside, P.M. 1997e. Greenhouse-gas emissions from Amazonian hydroelectric reservoirs: The example of Brazil's Tucuruí Dam as compared to fossil fuel alternatives. *Environmental Conservation* 24(1): 64-75.

Fearnside, P.M. 2000b. Global warming and tropical land use changes: greenhouse gas emissions from biomass burning, decomposition, and soils in forest conversion, shifting cultivation and secondary vegetation. Climatic Change 46(1/2):115-158.

Fearnside, P.M. nd. Biomass of Brazil's Amazonian forests. (in preparation).

Fearnside, P.M. and J. Ferraz. 1995. A conservation gap analysis of Brazil's Amazonian vegetation. *Conservation Biology* 9(5): 1134-1147.

Fearnside, P.M. and W.M. Guimaràes. 1996. Carbon uptake by secondary forests in Brazilian Amazonia. *Forest Ecology and Management* 80(1-3): 35-46.

Houghton, J.T., G.J. Jenkins and J.J. Ephraums (eds.). 1990. *Climate Change: The IPCC Scientific Assessment*. Cambridge University Press, Cambridge, UK 364 p.

Houghton, J.T., L.G. Meira Filho, B.A. Callander, N. Harris, A. Kattenberg and K. Maskell (eds.). 1996. *Climate Change 1995: The Science of Climate Change*. Cambridge University Press, Cambridge, UK 572 p.

Houghton, R.A. 1991. Tropical deforestation and atmospheric carbon dioxide. *Climatic Change* 19(1-2): 99-118.

Houghton, R.A., R.D. Boone, J.R. Fruchi, J.E. Hobbie, J.M. Melillo, C.A. Palm, B.J. Peterson, G.R. Shaver, G.M. Woodwell, B. Moore, D.L. Skole and N. Myers. 1987. The flux of carbon from terrestrial ecosystems to the atmosphere in 1980 due to changes in land use: Geographic distribution of the global flux. *Tellus* 39B: 122-139.

ISTOÉ. 1997. "A versào do Brasil" ISTOÉ [São Paulo] No.1463, 15 de outubro de 1997, p. 98.

Myers, N. 1989. *Deforestation Rates in Tropical Forests and their Climatic Implications*. Friends of the Earth, London, UK 116 p.

Myers, N. 1991. Tropical forests: Present status and future outlook. *Climatic Change* 19(1-2): 2-32.

Perry, H. and H.H. Landsberg. 1977. Projected world energy consumption. pp. 35-50 In: United States National Academy of Sciences (NAS) *Energy and Climate*. NAS Press, Washington, DC, USA.

Schimel, D. and 75 others. 1996. Radiative forcing of climate change. pp. 65-131 In: J.T. Houghton, L.G. Meira Filho, B.A. Callander, N. Harris, A. Kattenberg and K. Maskell (eds.) *Climate Change 1995: The Science of Climate Change*. Cambridge University Press, Cambridge, UK 572 p.

Shine, K.P., R.G. Derwent, D.J. Wuebbles and J-J. Morcrette. 1990. Radiative forcing of climate. pp. 41-68 In: Houghton, J.T., G.J. Jenkins and J.J. Ephraums (eds.) *Climate Change: The IPCC Scientific Assessment*. Cambridge University Press, Cambridge, UK 365 p.

Skole, D. and C. Tucker. 1993. Tropical deforestation and habitat fragmentation in the Amazon: Satellite data from 1978 to 1988. *Science* 260: 1905-1910.

Setzer, A.W., M.C. Pereira, A.C. Pereira Junior and S.A.O. Almeida. 1988. *Relatório de Atividades do Projeto IBDF-INPE "SEQE"-Ano 1987*. Pub. No. INPE-4534-RPE/565. Instituto de Pesquisas Espaciais (INPE), São José dos Campos, São Paulo, Brazil. 48 p.

Setzer, A.W. and M.C. Pereira. 1991. Amazonia biomass burnings in 1987 and an estimate of their tropospheric emissions. *Ambio* 20(1):19-22.

Traumann, T. 1998. Os novos vilóes: Ação dos sem-terra e de pequenos agricultures contribui para o desmatamento da Amazônia. *Veja* [São Paulo] (4 Feb. 1998) 31(5): 34-35.

Watson, R.T., H. Rodhe, H. Oeschger and U. Siegenthaler. 1990. Greenhouse gases and aerosols. pp. 1-40 In: J.T. Houghton, G.J. Jenkins and J.J. Ephraums (eds.) *Climate Change: The IPCC Scientific Assessment*. Cambridge University Press, Cambridge, UK 364 p.

WRI (World Resources Institute). 1990. *World Resources Report 1990-91*. World Resources Institute, Washington, DC, USA 383 p.

Integrating Ecological Complexity
into Economic Incentives
for Sustainable Use
of Amazonian Rainforests

James R. Kahn
Frank McCormick
Vicente de Paulo Queiroz Nogueira

James R. Kahn is Professor of Economics, and Director of Environmental Studies, Washington and Lee University, Lexington, VA 24450.

Frank McCormick is Professor, Department of Ecology and Evolutionary Biology, University of Tennessee.

Vicente de Paulo Queiroz Nogueira is Presidente, Instituto de Proteção Ambiental do Amazonas (Secretary of the Department of Environmental Protection for the State of Amazonas), and Professor of Centro do Ciências Ambiental, Universidade do Amazonas.

This paper was prepared for Dimensões Humanas Da Mudança Climática Global E Do Manejo Sustentável Das Florestas Das Américas: Uma Conferência Interamericana (The Human Dimensions of Global Climate Change and Sustainable Forest Management in the Americas), Brasília, December 1-3, 1997 and The Southern Economic Association Meetings, Atlanta, November 21-23, 1997.

This research was partially supported by the National Science Foundation's National Center for Environmental Decision Making Research, Joint Institute for Energy and Environment; and also by the Integrated Assessment of Global Change Program of the Office of Research, Department of Energy. The conclusions reached in this paper should not be attributed to the National Science Foundation, Department of Energy, or any of the institutions with which the authors are affiliated.

[Haworth co-indexing entry note]: "Integrating Ecological Complexity into Economic Incentives for Sustainable Use of Amazonian Rainforests." Kahn, James R., Frank McCormick, and Vicente de Paulo Queiroz Nogueira. Co-published simultaneously in *Journal of Sustainable Forestry* (Food Products Press, an imprint of The Haworth Press, Inc.) Vol. 12, No. 1/2, 2001, pp. 99-122; and: *Climate Change and Forest Management in the Western Hemisphere* (ed: Mohammed H. I. Dore) Food Products Press, an imprint of The Haworth Press, Inc., 2001, pp. 99-122. Single or multiple copies of this article are available for a fee from The Haworth Document Delivery Service [1-800-342-9678, 9:00 a.m. - 5:00 p.m. (EST). E-mail address: getinfo@haworthpressinc.com].

99

SUMMARY. Tropical rainforests are complex ecological systems capable of supporting human populations in a sustainable fashion. Unfortunately, much of current economic activity exploits the tropical rainforest in an unsustainable fashion, eliminating the ecological services which flow from the forest and reducing the future income earning potential of the forest inhabitants.

Over the last decade, much scientific research has taken place which examines the ecological characteristics of both the forest and exploitation activities which are necessary to ensure its resilience and ability to regenerate. However, less work has been done on how to translate the knowledge of these ecological parameters into practical and enforceable policy instruments which can constrain economic activity in a fashion which ensures the viability of the forest and thus the sustainability of the activity.

The paper reviews the scientific research in sustainable tropical forestry and highlights the ecological factors which have been shown to be important in maintaining forest productivity, resilience and sustainability. Alternative ways of incorporating these ecological factors into policy instruments are discussed, for both commercial forestry applications and for subsistence activities. *[Article copies available for a fee from The Haworth Document Delivery Service: 1-800-342-9678. E-mail address: <getinfo@haworthpressinc.com> Website: <http://www.haworthpressinc.com>]*

KEYWORDS. Rainforest, sustainable, ecological services, forest productivity, policy

INTRODUCTION

The environmental economics literature has defined and examined a number of economic incentives which are designed to achieve the target level of pollution at the minimum abatement cost to society. Taxes, marketable pollution permits, deposit-refund systems, performance bonding systems and liability systems have all been discussed, and shown to be preferable in many ways to traditional command and control methods, such as setting standards for individual polluters, or specifying abatement equipment or production technology.[1]

The research which analyzes the comparative properties of alternative policies for emissions reduction has reflected the societal concern with the reduction of pollution emissions, and has been ongoing for the past three decades. More recently, concern has shifted from the reduction of emissions in developed countries to sustainability and the protection of renewable resource systems in developing countries. While much insight can be gained from the earlier literature on the economics of emissions abatement, new insights must be developed in order to protect these environmental resources. First, it is important to understand both economic behavior and ecological behavior when developing policy instruments. Second, many of the environmental systems in danger of overexploitation and collapse are threatened by economic activities of the non-money subsistence sector of the economy. The environmental policies referenced above were designed to be implemented in a market economy and may not be as effective in the traditional subsistence sector.

The goal of this paper is to contribute to the literature by examining policies to promote the sustainable use of environmental resource systems. The formal sector of the economy's uses of renewable resource systems are studied through the examination of commercial forestry in Amazonas, Brazil. The informal sector of the economy's uses of renewable resource systems are analyzed through the examination of settler activities in the Mamirauá Sustainable Development Reserve in Amazonas.

RENEWABLE RESOURCES, ECOLOGICAL SERVICES AND THE ECONOMIC PROCESS

Before embarking on the discussion of both the economics of sustainable forestry and the policy instruments to achieve sustainable forestry, it is important to create some context for examining questions of sustainability. Much of the past literature in economics[2] has focused on the question of sustainability with dependence on exhaustible resources. For example, Hartwick develops the "Hartwick Rule" which suggests that sustainable growth of output is possible if the scarcity rents associated with the extraction of the exhaustible resource are reinvested in human-made capital. It should be noted, that this result depends on an assumption of substitutability between natural capital and human-made capital.

The assumption of a high degree of substitutability between natural capital and human-made capital is not a bad assumption if natural capital is viewed to consist solely of exhaustible resources such as fossil fuels or mineral ores. For example, a reduction in the availability of oil can be compensated for by the development of more efficient engines or photovoltaic cells.

However, natural capital should be divided into two categories, exhaustible resources and environmental resources, which are fundamentally different in their degree of substitutability with human-made capital and human capital. Environmental resources consist of ecological systems which provide ecological services such as maintenance of global climate, biodiversity, carbon sequestration, oxygen production, maintenance of hydrological cycles, soil formation, nitrogen fixation in soil, nutrient cycling, primary productivity and so on. The services emanating from human-made capital are poor substitutes for the ecological services arising from environmental resources. A preliminary conclusion which can be reached from this brief conceptual discussion is that sustainable development paths are those that are conservative of environmental resources and protective of ecological services, and unsustainable development paths are those that are consumptive or destructive of environmental resources, thereby reducing the flow of ecological services.[3]

It is this approach to sustainability which will guide the discussion of sustainable forestry in this paper. Forestry practices which maintain the ability of the forest to continue to provide ecological services (including the provision of wood) are considered sustainable, but forestry practices which eliminate or reduce the flow of ecological services are considered unsustainable, even though they may provide a continual flow of wood and wood products. Under this definition, the conversion of undisturbed tropical forest to tree plantations would be considered unsustainable.[4]

ECOLOGICAL DETERMINANTS
OF TROPICAL FOREST SUSTAINABILITY

For development to be sustainable, planning and implementation should meet ecological, economic and social criteria. This section of the paper focuses on ecological determinants of tropical forest sustainability, while later sections of the paper examine how these ecological determinants, in combination with economic and social factors, can be used to shape policy instruments to help achieve sustainability.

Sustainable development of tropical rainforests should focus upon two fundamental ecological determinants: (1) severity of disturbance, and (2) capability to recover. Both determinants are influenced by extrinsic and intrinsic factors operating as exogenous and endogenous variables. Variables include spatial scale of disturbance, intensity of disturbance, intensity of other stressors at the time of (or preceding the disturbance), age and health of the rainforest ecosystem, and the degree to which human disturbances mimic natural disturbances.

Probability of recovery and time required are directly proportional to the severity of the disturbance. Extended recovery time contributes to secondary and delayed impacts such as soil erosion or compaction, nutrient or water depletion, loss of mycorrhizae and extreme temperature and light regimes. Severely altered micro environments favor invasions of exotic species which may significantly alter paths of rainforest recovery. Severity of disturbance is a measure of how extensively biotic and abiotic components are altered. If disturbance does not extend to extensive alteration of geological substrate and soils, residual populations of plants, animals, microbes and humans may survive in micro habitats. Recovery by residual populations is relatively rapid, predictable and dependent upon intrinsic ecological properties. If the disturbance is so sever that micro habitats are destroyed and colonization is required from distant sources, recovery is slow and stochastic (or even chaotic). If the disturbance is sufficiently intensive, or extensive, to diminish populations of reproductively mature individuals, pollinators or seed dispersers, ecosystem recovery can also be significantly altered or delayed.

The degree to which human disturbance mimics natural disturbances is the single most significant determinant of ecosystem recovery. Natural stressors such as wind, hurricanes, disease, old age and extreme weather events contribute to tree mortality. Dead and weak trees fall to the forest floor as frequently as one tree per hectare per year. Tree falls create gaps ranging from 30-50 meters in width. Various tree species are pre-adapted to invade gaps of different sizes. This "intermediate" level of natural disturbance is beneficial and necessary to maintain high species biodiversity and ecosystem health. Rainforest species are also well adapted to larger disturbances such as floods, fire, drought or hurricanes if the disturbances are rhythmic, cyclic or periodic in occurrence. Strategies for sustainable tropical forestry should be based upon an understanding of these types of ecosystem responses to various scales and intensities of disturbance. These strategies require long range planning based upon principles of ecological science and field experiments which transform theory into practice. In the absence of long range planning, extraction (disturbance) of forest products is too often episodic, or unnatural scale, and influenced by market failure.

There are several types of market failure which may be important in the case of forestry management. First, there is the ubiquitous problem of imperfect information, in which case logging firms or households do not know the appropriate technology for sustainable forestry. Second, there is the problem of insecure property rights. If people do not have long-term security in the property rights of their land, there is no incentive to engage in long-term optimization. Related to the problem of the insecurity of property rights is the granting of short-term timbering leases by either private or public owners of forest land. Short-term leases do not provide an incentive for the harvesting firm to treat the forest gently and conform to the ecological principles which are described above.

One may ask why land owners voluntarily agree to short-term leases, even though these lead to degradation of the forest and a loss in long-term income. The answer is that both public and private owners of forests may face constraints which force them to make decisions to maximize short-run economic gain, rather the long-term economic gain.

Governments may be forced to focus on short run economic gains because of immediate needs for income. This may be because of macroeconomic problems such as large external debt which requires servicing, or because of pressing needs to maintain or increase current consumption.[5]

Individual landowners will, in general have a greater rate of time preference than society as a whole, for both the usual reasons,[6] and because of short-term concerns with feeding their families. In particular, it is likely that the income paths of unsustainable and sustainable forestry take the shapes depicted in Figure 1. If unsustainable forestry has higher initial levels of income than sustainable forestry, and if credit is unavailable to small landowners, then they have an incentive to engage in unsustainable forestry or to write short-term harvesting contracts with forestry firms.

Of course, there is also a general market failure because of the global public good nature of tropical forests. Even if the within-country market failures could be addressed by policy, harvesting is likely to

FIGURE 1. Alternative Time Paths of Forestry Income

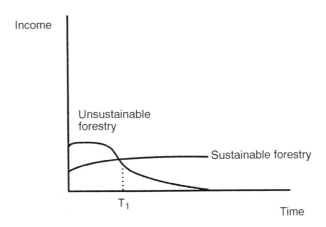

be less sustainable than optimal (from a global point of view) because the global benefits of forest preservation can not be captured by local landowners without an international agreement to transfer resources based on forest preservation.

TWO MODELS OF SUSTAINABLE FORESTRY

Unfortunately sustainability (McCormick, 1998) has not been a prerequisite for the majority of development projects undertaken in the Amazon. However, over the past two decades sustainable forestry has become the focal point of a steadily growing number of development projects in the Amazon. We offer three models of sustainable forestry in the Amazon. The first of these, "selective harvesting" involves the harvest of widely separated individual trees of numerous species. Disturbance is held to a minimum, enhancing natural forest regeneration. A very different approach, "the strip method," involves clear cutting long narrow areas in the midst of larger undisturbed areas of forest. The dimensions of these long fingers of clear-cut are intended to mimic those resulting from natural disturbances. If the mimicking is successful, natural ecological processes can be expected to contribute to regeneration of the forest. The "strip shelter belt" method contains elements of each of the preceding methods. Procedures are designed to mimic natural disturbance, minimize anthropogenic disturbance and to enhance natural regeneration.

The selective logging method is heavily dependent upon scientific research and technology to guide selection of individual trees for harvest. The strip method is heavily dependant upon the ecological theory of gap dynamics and intermediate disturbance hypotheses. The stip shelter belt method shares dependencies of both the other methods.

THE SELECTIVE LOGGING METHOD
OF SUSTAINABLE TROPICAL FORESTRY

The selective logging method of sustainable forestry requires a significant amount of scientific knowledge to conduct properly. A good example of the selective logging method can be found in the activities of Precious Woods, Ltd., a Swiss owned-Brazilian operated sustainable forest management.

Precious Woods owns 81,000 hectares of primary forest located approximately 250 km east of Manaus (capital city of the state of Amazonas) and 40 km west of Itacoatiara, an Amazon river town of 80,000 people. The most distinguishing feature of this project is the extent of research, forest inventory and log range planning. Management plans focus upon minimizing damage to forest understory and the area to be harvested, and upon maximizing biodiversity and sustainability. Approximately 25,000 hectares (30%) are placed in permanent reserve to serve as a source of seeds, pollinators, seed dispersers and wildlife, as well as providing baselines against which other sites may be compared. Approximately 5,000 hectares (6%) are dedicated to infrastructure such as roads and the mill site. Each year, a different work unit of approximately 2,000 hectares (2%) is selectively harvested, on an expected rotation cycle of approximately 25 years. Thus, at anyone time, 91% of the rainforest is in permanent or temporary reserve. The Celos Management System is based on research at Waginengen Agricultural University and was field tested in experimental plots in Suriname. Pre- and post-harvest plots (100 m^2) are established in each work unit and are monitored to estimate the severity of disturbance during extraction and recovery thereafter. Prior to extraction, all trees greater than 50 cm (dbh) are identified, measured and located on a geographic information system (GIS). As work progresses, smaller trees will be similarly inventoried and data entered into forest growth models to estimate available wood volume of each species at future dates. Traditional forest practices in the region have typically only utilized one to four species of trees. In contrast, this sustainable forestry operation extracts 40 species of trees (with an eventual goal of 60 species) to minimize the area disturbed annually and to avoid high-grading and associated reduction of biodiversity.

Specific procedures include the following:

1. Access into work units and extraction of trees is confined to trails 3.5 m in width and 80 m long which are located at 100 m intervals. Accordingly, no tree is more than 50 m from a trail. Disturbance of the forest understory is minimized.
2. Prior to cutting, all vines are removed in order to minimize the collateral damage to other trees and to minimize gap size.
3. Direction of tree felling avoids trees which are of economic or ecological value (such as palms which provide food and shelter for many plant and animal species).
4. No more than 80% or reproductively mature individuals are removed from a work unit in order to insure a nearby source of seed for recruitment of new individuals and to provide a continuing food source for dependent animal species.
5. Only 10% (40 m^3) of wood volume per hectare is removed, to ensure that nutrient and carbon loss are within sustainable limits.
6. Track skidders used to extract trees are restricted to trails. Logs are trimmed and transported to trails by cable in order to minimize disturbance.
7. Prior to extraction by cable, logs are cut to 7 m lengths to minimize sway during extraction. This further reduces disturbance.
8. Once logs are on the trail, they are removed by wheeled vehicles in order to minimize disturbance of trails.

Economic theory suggests that long-term sole ownership of an open-access renewable resource should lead to socially efficient outcomes.[7] Of course, a market failure of imperfect information could always lead to sub-optimal outcomes, but in the case of Precious Woods, they have developed the scientific information necessary to ensure a sustainable operation of the forest. However, despite this good information on the production side, a market failure of imperfect information still threatens the achievement of a socially optimal outcome. The imperfect information is on the consumer side of the market. The primary information problem is that consumers of wood (either as an intermediate or final good) are only familiar with a very few species of wood, focusing primarily on well known woods such as mahogany and rosewood. Even though many of the other 40 to 60 species of trees that can be harvested from this section of the forest are just as beautiful and durable as mahogany and rosewood, the demand for these woods is much lower, simply because people do not know about them.

Green and Kahn (1997) talk about the importance of demand-side policies for sustainable management of renewable resource systems. Although the paper by Green and Kahn uses fishery resources as an example, the central point of the paper is directly transferable to forests. This point is the renewable resource literature focuses on maximizing efficiency, given a demand curve. However, as Smith (1968) points out, under these conditions, the extinction of the harvested species can be optimal. Increasing the cost of production (such as through taxes, individual transferable quotas, or inefficient policies such as direct controls) is one way of addressing this problem. However, if demand is sufficiently strong, the stock will eventually be harvested to zero.

The policy recommendation resulting from Green and Kahn's paper, is that problems that are demand in origin must have a solution that is founded in demand. For problems of excess depletion of wild fishery stocks, policies aimed at the stimulation of a substitute good through the production of aquacultured fish is suggested. The problem at hand for selective sustainable forestry is related, but has slightly different dimensions. The problem is that there is too much extraction of high profile species and insufficiently high demand for alternative species, because consumers are unaware of the properties of these other types of wood.

In this case, information must be supplied to consumers (both domestic and international) about the properties of these alternative woods, creating a market for the many other species. Also since consumers in general care about the preservation of the Amazonian rainforest, a certification program must be developed to distinguish wood produced in a sustainable fashion from wood produced in an unsustainable fashion. Presumably, there is a higher willingness to pay for sustainable forest products, which would encourage the development and adoption of sustainable forestry practices. Despite the best use of science and sustainable forestry practices, the long-term sustainability of the forest and forest industries is in doubt until these demand-related issues are addressed.

THE STRIP METHOD
OF SUSTAINABLE TROPICAL FORESTRY

The selective harvesting method of forestry requires a significant investment in information in order to implement it properly. As discussed above, its implementation requires a complete mapping of trees, and the implementation of an optimization program, with constraints on the distribution of trees, temporal distribution of fruit, presence of pollinators and seed dispersers, and other ecological factors. Because of this, it is also a method for which it is more difficult to develop encouraging policy instruments, because the implementation, monitoring and enforcement of the instruments would have correspondingly high information requirements for the regulatory agency.

In contrast, the strip method requires less information, as it is based on clear-cutting, but of small areas. The disturbances associated with harvesting can most closely mimic natural disturbances if the clear cut areas are appropriately configured. It is even possible that this type of activity could be monitored with remote sensing techniques, such as satellite imagery or aerial overflights of downward-looking radar. If operational measures of the appropriate configurations are defined and can be monitored, a deposit-refund or performance bonding system can be developed to encourage this type of sustainable forestry.

First, as the research from a joint INPA/Smithsonian Institute project[8] indicates, disturbed areas should be small in comparison to the undisturbed area. As Thomas Lovejoy indicated,[9] the only workable configuration is "islands of development in a sea of forest." Therefore, one operational measure of the recoverability of a harvested forest would be the ratio of undisturbed area to disturbed area. This is illustrated in panels A and B of Figure 2. The higher the ratio of undisturbed area to disturbed area (hereafter referred to as the undisturbed area ratio), the greater the recoverability of the forest. This is because lower undisturbed area ratios will result in loss of many of the ecological characteristics outline in section three. In particular, as the undisturbed area ratio shrinks, animals which fulfil the roles of both pollinators and seed dispersers leave the area, which drastically inhibits the return of the forest to its original state.

FIGURE 2. Developing Operational Measures of Forest Sustainability

Panel A: Low ratio of undisturbed area to disturbed area

Panel B: High ratio of undisturbed area to disturbed area

Panel C: Low ratio of edge of disturbed area to surface area of disturbed area

Panel D: High ratio of edge of disturbed area to surface area of disturbed area

 Undisturbed Forest

 Disturbed area of harvesting activity

A second operational measure of the recoverability of the forest is the ratio of the edge of the disturbed area to the surface area of the disturbed area (hereafter referred to as the edge ratio), as depicted in Panels C and D of Figure 2. Areas of high edge ratio are long-narrow strips which mimic the clearings created by the natural falling of emergent trees. Since many of the regenerative processes operate at the edge of the clearing, the greater the edge ratio, the more potential for these processes to operate. Also, with lower edge ratios, the cleared area is more square or circular in configuration (Panel C) and more soil is exposed to the sun and the negative effects of compaction, erosion, and loss of leaf liter, mycorrhizae and nutrients.

The strip shelter belt method combines the characteristics of the strip method and the selective logging method, and can be interpreted as an extension of the strip method. The Central Selva Resource Management (CSRM) Project (Tosi 1981, Hartshorn 1981) in the Pal Cazu Valley of the Peruvian Amazon is an excellent example of the commercial implementation of these theories of sustainable forestry, providing a field test of the importance of mimicking these natural disturbances through the edge/surface area and undisturbed/undisturbed area factors.[10]

If these factors are important in the determination of forest recoverability, then they can provide the basis of compliance standards. In other words, standards can be defined in terms of desired levels of the undisturbed area ratio and the edge ratio. These standards can then be used in conjunction with either direct controls or economic incentives. However, given the problems described above in terms of short-run perspectives, economic incentives might be more desirable. In particular, economic incentives (such as performance bonding or deposit-refund systems) which collect the penalty for noncompliance *before* the harvesting activities take place may be more preferable. Harvesting companies which leave the forest in condition to recover (according to the operational measures defined above) would then be returned all or a portion of their bond or deposit.

This is very similar to the process used to ensure reclamation of areas which are strip mined in the United States. Mining companies are required to post a bond and if they do not reclaim the land and restore its contours and vegetative cover, the bond is forfeited and the money is used to pay a third party to reclaim the land. As long as the magnitude of the bond is greater than the cost of reclamation, reclamation will take place.

The situation is slightly different in the case of harvesting of a tropical forest, because the forest itself is the source of reclamation activity, rather than human engineering. Instead of trying to provide incentives for synthetic restoration, the forest performance bonding system seeks to provide incentives to harvest the forest in such a way that the forest retains its recuperative capabilities. Thus, the performance bond must exceed the opportunity cost of meeting the sustainability standards defined above.

Different methods can be defined for relating the refund to the standards. For the time being, we will assume that there is only one standard, and the problem of dealing with two standards simultaneously (both the undisturbed area ratio and the edge ratio) will be dealt with shortly.

The easiest way to relate the operational measure of sustainability to the refund is to define an ideal or target level of the measure, and return the performance bond in proportion to the achievement of the ideal level. For example, in Figure 3, the actual level of the operational measure is measured on the horizontal axis. If M_0 represents the ideal level, then the performance bond would be returned according to the function in Figure 3. In other words, if the actual measure is 40% of the ideal measure, then 40% of the deposit is returned.

While this proportionate system is easy to define, it is very naive as it assumes a linear relationship between the operational measures and the recoverability of the forest. It is rather unlikely that this type of relationship is linear, and it is quite likely that there are threshold effects.

Figure 4 shows a refund function based on a declining marginal product of the operational measure, but no threshold. This would be the appropriate shape of a refund function if, as one approached the ideal level of the operational measure, the contribution to recoverability of another unit of the operational measure became smaller and smaller. This would indicate that the difference between 80% and 90% of the ideal measure is bigger than the difference 90% and 100% of the ideal measure. In contrast Figure 5 shows a refund function based on an increasing marginal product of the operational measure. Although we believe it likely that the marginal product is decreasing as one

FIGURE 3. Refund Function (Constant Marginal Product, No Threshold)

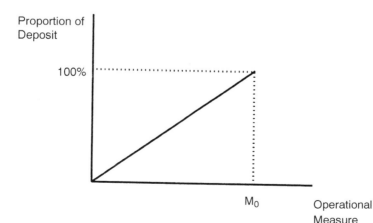

FIGURE 4. Refund Function (Declining Marginal Product, No Threshold)

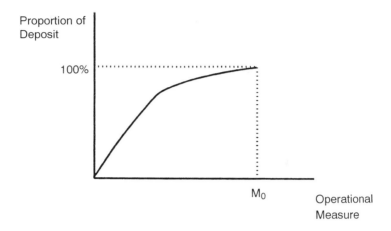

FIGURE 5. Refund Function (Increasing Marginal Product, No Threshold)

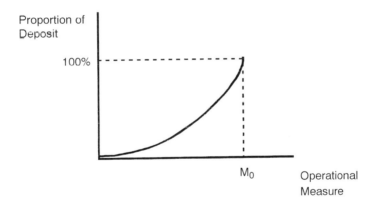

approaches the ideal measure, more research needs to be conducted to show whether it is increasing or decreasing, and the functional form may be a related to soil type, species mix, topography and other local factors.

It is also likely that a threshold exists such that if the operational measures drop below the threshold level, recovery becomes doubtful.[11] The easiest way to incorporate such as ecological threshold into policy would be to return all of the performance bond if the operational measure is above the threshold level, and none of the performance bond if the operational measure is below the threshold level. However, such a simple system does not give harvesting firms an incentive to perform better than the threshold level and move towards the ideal level. The refund functions depicted in Figures 6 and 7 provide such an incentive. Again, we view the declining marginal product function of Figure 6 as more conforming with reality, but more research needs to be done before this hypothesis can be accepted.

FIGURE 6. Refund Function (Decreasing Marginal Product with Threshold)

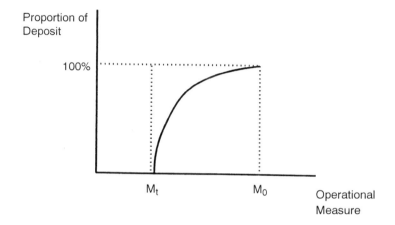

FIGURE 7. Refund Function (Increasing Marginal Product with Threshold)

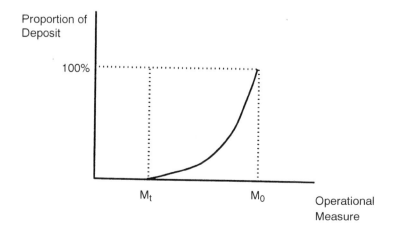

Finally, it is important to develop a refund function which incorporates multiple operational measures of recoverability. In this case, we have defined two operational measures, the undisturbed area ratio and the edge ratio. The easiest way to combine the two into one system is to average them, with a provision that if either of the measures is below the threshold level, then no refund is returned. However, if there are synergies between the two measures, then the combining function needs to be a nonlinear function. For example, the total refund can be the product of the two individual refund functions. Again, more research needs to be conducted to determine the most appropriate functional forms if it is true that there are important synergies between the undisturbed area ratio and the edge.

CURRENT POLICY IN THE STATE OF AMAZONAS

Amazonas is a state in northwestern Brazil (see Figure 8), and considered a "core" Amazon state, all the other states which border it

FIGURE 8. The State of Amazonas, Brazil

(including states in other countries) are also part of the Amazon region. This proves to be a very important geographic consideration, as it is the "periphery" states further to the south and east (such as Acre, Rondonia, Mato Grosso, and Tocantins) where the more pronounced deforestation is taking place. In fact, Amazonas is still 97% in its original forest cover.

Amazonas is also different from many other Brazilian states to the extent that the state government is active in environmental protection. Environmental quality is regulated primarily by direct controls through the issuance of environmental licenses by the state environmental agency, the Instituto de Proteção Ambiental do Amazonas (IPAAM). Initially, IPAAM focused on regulating the manufacturing activities of the Manaus Free Trade Zone and on mining. More recently, increasing attention has been placed on industrial and commercial uses of renewable resources, with regulations in place for forestry and being developed for fisheries.

As mentioned above, IPAAM relies primarily on direct controls. It is not surprising to find this "government intensive" method for maintaining environmental quality in Amazonas, as there is a high degree of government involvement in the economy as a whole. The Manaus Free Trade Zone is a program of fiscal incentives designed to encourage manufacturing and the industrial sector. It is highly unlikely that the large manufacturing center which is currently located in Manaus would have developed without this heavy government subsidization, although the manufacturing center may have now developed the critical mass to exist without the subsidies.

Environmental protection in Amazonas is generated by requiring industrial activities (including commercial forestry) to file an environmental management plan with IPAAM, and if the plan passes muster, the activity is granted a license, which incorporates the provisions of the plan. If a firm operates without a license, or violates the provisions of its license, it is subjected to heavy penalties, which could include the loss of fiscal incentives, environmental fines, or even closure of the operation.

Industrial forestry activities are controlled in the same fashion. For example, Precious Woods has filed an environmental management plan with IPAAM, and IPAAM inspectors have approved the plan and periodically inspect Precious Woods' operations to make sure that they are complying with the provisions of the plan.

Recently, a large number of Malaysian and other Asian forestry concerns have either purchased land in Amazonas or entered into joint ventures with Brazilian firms. Given the environmental destruction which has followed these firms in Southeast Asia, the state government developed concerns that these firms may practice unsustainable forestry in Amazonas. The fear was that the corporations would file sustainable forestry plans with IPAAM, but then harvest in an unsustainable fashion, and disappear in a corporate mist of phantom holding companies and other financial camouflage. It should be noted that the state government lacks the constitutional authority to prohibit the sale of land to Malaysian firms.

Obviously, this situation is exactly the type that performance bonding was developed to control. The system which was instituted in Amazonas most apply equally to all forestry operations, and can not be limited to Malaysian firms or foreign owned firms.

The implemented system is slightly different than that which is proposed earlier in this paper. First, rather than linking the system directly to an operational measure such as that which was proposed above, each firm is required to file and obtain approval for a sustainable harvesting management plan. Once operations have begun, the operations are regularly inspected and if there is observed deviation from the approved plan, then the companies are fined, must correct the discrepancy and then post a performance bond before further operations take place. Further deviations from the management plan would result in the forfeiture of the performance bond and permanent closure. In terms of the discussion in the preceding section of "ideal levels" and "thresholds," the management plan could be viewed as defining a threshold (plus a margin of safety), and the performance bond is forfeited if the threshold is not attained.

It might seem unusual to set up a performance bonding system which is not implemented until after some environmental harm takes place. This "one-strike rule" was developed because a universal performance bonding system would have placed high costs on local forestry firms which were operating with acceptable practices. Imposing this high cost on local firms who were practicing good forestry was not viewed to be politically acceptable.

The first violation of this system recently occurred, when a Malaysian firm bought wood illegally obtained from an Indian reservation. The firm was fined, closed down and can not re-open without posting the bond.

FOREST SUSTAINABILITY
AND THE SUBSISTENCE SECTOR

In many ways, the industrial sector of the economy represents a much easier sector for which to develop environmental policy than the subsistence sector. Industrial activities are identifiable, and more easy to monitor and enforce penalties upon than the subsistence sector. In particular, because the participants in the subsistence sector have no accumulation of capital, environmental penalties such as fines become meaningless. Even if monitoring is possible and the environmental degradation can be attributed to particular individuals, it is difficult to fine someone who has no money. Environmental taxes do not work in the non-cash economy, and subsistence farmers have no money with which to participate in a deposit refund or performance bonding system. Even the threat to move the offender off his or her land will not lead to effective environmental policy, because the family will simply relocate somewhere else and practice environmental unsound farming and forestry techniques there (or move to the city and exacerbate the environmental and social problems found in the large cities).

Before discussing the role of policy in generating sustainability of subsistence activities, it is necessary to distinguish between three distinct groups of people engaged in subsistence activities. These groups are indigenous peoples, caboclos and recent immigrants. Indigenous peoples are the native Indian tribes which lived in the Amazon region before the period of European colonialization, caboclos are settlers of mixed European and Indian descent who have lived in the Amazon region for many generations. Recent immigrants are the people who have immigrated from the non-Amazonian regions of Brazil, such as the large cities of the south. Recent migrants generally employ production techniques which have not co-evolved with the environment[12] and are unsustainable and destructive of the forest and other environmental resources. Indigenous peoples and caboclos have generally developed production techniques which are in harmony with the environment, and which have proven to be sustainable over many generations.

Recent social, economic and demographic changes have eroded the sustainability of the indigenous and caboclo economic activities. Increased populations, reduced availability of land, increased economic interchange with the industrialized urban centers, the globalization of consumerism, and other factors have increased the need for higher yields from the renewable resource systems. For example, the increased manufacturing of electronic goods in Manaus has increased local demand for wood from which to make corrugated cardboard packaging. This increased demand, accompanied by increased need for cash income among caboclo families, has lead to increased harvesting of timber among caboclo families.

As stated above, these social and economic changes threaten the sustainability of the forest, but it is difficult to develop policies to regulate the subsistence sector. The important policy question is whether it is possible to develop the scientific information necessary to increase income from sustainable uses of the forest, without jeopardizing the long-term viability of the forest or the communities. Following this, how is the scientific information communicated to the individual communities so that they can improve their quality of life without threatening ecological resources.

The Mamirauá Sustainable Development Reserve (MSDR) is a model for promoting sustainability in the rural subsistence sector. The area was identified in the early 1980s as an area of great ecological significance, and is particularly noted as habitat for many threatened species, including a rare primate, the white ouakari. The ouakari is not hunted, but is threatened by loss of habitat through timbering activity. The MSDR is also part of the Central Amazonia protected area corridor being developed by the G7 Parks and Reserves Project.

The unusual feature of MSDR is that the people who have traditionally lived in the area, are not being expelled as part of the development of the sanctuary. The people have co-existed with the forest for many generations, yet current social and economic trends endanger the forest. The focus of the project is scientific research to help the resident population improve their standard of living, while maintaining the quality of the environmental resources.

For example, one focus of research has been on fishery biology. Increased knowledge of the reproductive, feeding and migratory habits of important food fishes can be used to modify fishing habits to increase long-term yield. For example, by suggesting harvest restrictions in certain areas during certain periods, suggesting the targeting of certain sizes of fish and other measures, it was expected that fish populations and sustainable harvests of fish could become larger. Again, there is the potential problem here of requiring short-term sacrifice for long-term benefit (see Figure 1) and regulations based on this type of scientific information have been notoriously unsuccessful throughout the world, as evidenced by the world fisheries crisis and the collapse of many fishery of both global and local significance. This is particularly true in the United States, where commercial fishers aggressively fight against all types of regulation and attempt to discredit the scientific research upon which fishery regulations are based.

Given the problems of implementing direct controls or economic incentives in a subsistence economy, and given the global lack of success with fishery regulations, the State of Amazonas opted for a cooperative education rather than a regulatory approach to the problem. Scientists and project managers worked hard to integrate the local residents into the project, involving them in decision making and creating a situation where the local residents believed that the scientists were playing on their team, rather than against them. This gave the scientists credibility in the eyes of the caboclos, and made them willing to integrate the scientific findings into their economic production techniques. When fishery yields began to increase after one or two seasons, this gave the scientists and the project even more credibility and will increase the likelihood that additional scientific findings related to forestry and agroforestry will be implemented by the local residents.

CONCLUSIONS

Although the rainforest is characterized by tremendous ecological complexity, it is possible to develop policy instruments which reflect this ecological complexity. For commercial harvesting or timber it is possible to develop operational measure of the potential recoverability of the forest such as the ratio of the area of the undisturbed area to the disturbed area, and the ratio of the edge of the disturbed area to the surface area of the disturbed areas. A performance bonding or deposit-refund system can be developed which returns the bond or deposit based on the harvesting firms performance with regard to their operational measures.

The subsistence sector generates additional difficulty in the development of policy, because of barriers to the successful implementation of both direct controls and economic incentives. The development of scientific knowledge which can improve the yield of sustainable production techniques, along with stakeholder involvement and an effective education process can lead to the adoption of these sustainable techniques among subsistence forest dwellers.

Although this paper discusses policies which have significant potential to contribute to the widespread adoption of sustainable uses of the rainforest, much more research needs to be done to assess their potential. In particular, the policies need to be implemented in different locations to see how people in different situations react to the policies.

NOTES

1. Of course there are exceptions to the general preferability of economic incentives over direct controls. For an advanced treatment of the relative merits of the two alternative approaches to pollution control see Baumol and Oates. Undergraduate textbooks which also discuss these issues include Tietenberg (1997) and Kahn (1998).

2. In the modern period, this literature begins with the work of Barnett and Morse (1967) and extends through the work of Toman et al. (1994) and Hartwick (1995).

3. More discussion of the role of ecological services in the economic process can be found in Kahn and O'Neill (1997a and 1997b).

4. This should not be taken to imply that forest plantations are always a bad idea. In fact, the conversion of clear cut barren areas or unproductive forest plantation increases ecological services, and at the same time reduces the need for forest products from primary forest.

5. See Kahn and McDonald (1994 and 1995) for a discussion of how external debt and other economic pressures can cause countries to engage in more myopic economic strategies.

6. There are a number of arguments why the social rate of time preference should be higher than individual rates of time preference. These arguments include the ability of society both to invest in a more diversified portfolio of investments and spread the risk over more individuals. In addition, there are ethical reasons for a higher social rate of time preference including responsibility to future generations.

7. See Gordon (1954) and Scott (1955), for example.

8. This work examines the critical size of tropical forest ecosystems. See Lovejoy et al. (1986b).

9. Presentation at the Environmental Sciences Division, Oak Ridge National Laboratory, 1993. Also see Lovejoy (1986a).

10. This project illustrated the importance of public participation, education and other social factors in developing sustainable economic activities, but as important as these factors are, they are beyond the scope of this paper.

11. The threshold can also be defined at a level below which recovery can still take place, but occurs at an unacceptably low level.

12. See Norgaard for a discussion of environmental and economic co-evolution.

REFERENCES

Baumol, W.J. and W.E. Oates (1988). *Theory of Environmental Policy*, London: Cambridge University Press.

Gordon, H.S. (1954). Economic Theory of a Common Property Resource: The Fishery. *Journal of Political Economy* 62: 124-42.

Green, Trellis and James R. Kahn (1997). Demand-Side Approaches to Mitigating Open-Access Externalities: The Case of Aquaculture and the Wild Fishery, Unpublished paper, University of Tennessee.

Hartshorn, G. (1981). Forestry Potential in the Pal Cazu Valley, In *Central Selva Resource Management*, JRB Associates, Inc., McLean, VA, USA.

Hartwick, John M. (1995). Constant Consumption Paths in Open Economies with Exhaustible Resources. *Review-of-International Economics* 3(3): 275-83.

Kahn, James R. (1998). *The Economic Approach to Environmental and Natural Resources*, 2nd edition, Fort Worth: Dryden Press.

Kahn, James R. and Judith A. McDonald (1995). Third World Debt and Tropical Deforestation. *Ecological Economics* 12: 107-123.

Kahn, James R. and Judith A. McDonald (1994). Investigating the Linkages Between Debt and Deforestation in *The Causes of Tropical Deforestation*, edited by Katrina Brown and David Pearce, University College London Press.

Kahn, James R. and Robert V. O'Neill (1997). Ecological Stability as a Source of Economic Scarcity presented at the Canadian Society for Ecological Economics, Hamilton, Ontario.

Kahn, James R. and Robert V. O'Neill (1999). Ecological Interaction as a Source of Economic Irreversibility. *Southern Economic Journal* 66(2): 381-402.

McCormick, J.F. (1998). Principles of ecosystem management and sustainable development. In *Ecosystem Management for Sustainability*, JD. Peine, ed. St. Lucie Press.

Lovejoy, Thomas (1986a). Conservation Planning in a Checkerboard World: The Problem of the Size of Natural Areas. In *Land and Its Uses, Actual and Potential: An Environmental Appraisal*, Last, F.T., Hotz, M.C.B. and B.G. (eds.), pp. 289-301.

Lovejoy, Thomas, et al. (1986b). Edge and Other Effects of Isolation on Amazon Forest Fragments. In *Conservation Biology: The Science of Scarcity and Diversity*, Soule, M.E. (ed.), pp. 257-285.

Scott, A.D. (1955). The Fishery: The Objectives of Sole Ownership, *Journal of Political Economy* 63: 116-24.

Smith, V.L. (1968). Economics of Production for Natural Resources. *American Economic Review* 58: 409-431.

Tietenberg, Tom, (1997). *Environmental Economics and Policy*. New York: Addison Wessley

Toman, Michael. John Pezzey and Jeffrey Krautkraemer (1994). Neoclassical Economic Growth Theory and "Sustainability," Resources for the Future, Energy and Natural Resources Division Discussion Paper: ENR93-14.

Tosi, Joseph A. Jr. (1981). Forestry Potential in the Pal Cazu Valley, In *Central Selva Resource Management*, JRB Associates, Inc., McLean, VA, USA.

PART III:
NORTH AMERICAN FORESTS

The Carbon Cycle
and the Value of Canadian Forests

Mohammed H. I. Dore
Mark Johnston

SUMMARY. The model derives an equation of the value of forests from a dynamic optimizing framework. This equation gives the marginal social opportunity cost (MSOC) value of the forests, with value

Mohammed H. I. Dore is Professor of Economics, Brock University, St. Catharines, Ontario, Canada, L2T 2M5 (E-mail: dore@adam.econ.brockU.ca).

Mark Johnston is affiliated with Forest Ecosystems Branch, Saskatchewan Environment and Resource Management, Box 3003, Prince Albert, SK, Canada S6V 6G1 (E-mail: johnston@derm.gov.sk.ca).

This paper was first presented at The Annual Conference of Canadian Resource and Environmental Economics, Simon Fraser University, Vancouver, B.C., September 30-October 1, 1995. The authors are grateful to Dr. Bill Veloce (Brock University) and Dr. Patricia Roberts-Pichette (Environment Canada) for helpful comments on an earlier version, but the usual disclaimer applies. This research is funded by the Social Sciences and Humanities Research Council of Canada, grant # 410-94-1121.

[Haworth co-indexing entry note]: "The Carbon Cycle and the Value of Canadian Forests." Dore, Mohammed H. I., and Mark Johnston. Co-published simultaneously in *Journal of Sustainable Forestry* (Food Products Press, an imprint of The Haworth Press, Inc.) Vol. 12, No. 1/2, 2001, pp. 123-151; and: *Climate Change and Forest Management in the Western Hemisphere* (ed: Mohammed H. I. Dore) Food Products Press, an imprint of The Haworth Press, Inc., 2001, pp. 123-151. Single or multiple copies of this article are available for a fee from The Haworth Document Delivery Service [1-800-342-9678, 9:00 a.m. - 5:00 p.m. (EST). E-mail address: getinfo@haworthpressinc.com].

123

added as the *numeraire*. However, the shadow price is subject to the restrictive complementary slackness condition, which yields a zero shadow price whenever a constraint in not fully binding. To overcome this problem, we assume that value-added is valued at par only when there is no externality. In this formulation the shadow price is a cardinal index of the relative scarcity of the forests.

The econometric estimates of the shadow price of the forests and the resulting MSOC value of forests per hectare are obtained by a time trend model (Model 2) and an ARIMA model (Model 3). We regard the ARIMA model estimates to more representative of current conditions, and the estimated shadow price of the forests (in 1990) is a premium of the order of 25 to 32 percent. The corresponding MSOC value of a hectare of forest is between $350 to $412 in 1986 constant dollars. Naturally the estimates are subject to the limitations of the data. *[Article copies available for a fee from The Haworth Document Delivery Service: 1-800-342-9678. E-mail address: <getinfo@haworthpressinc.com> Website: <http://www.haworthpressinc.com>]*

KEYWORDS. Carbon cycle, value of forests, dynamic optimization, marginal social opportunity test, econometric estimates

INTRODUCTION

In the theory of renewable and non-renewable natural resources, what is the appropriate concept of scarcity rent? How is this rent likely to change over time? A review of literature on this subject shows that a number of economists have divergent views. Brown and Field (1978) and Fisher (1979) argue that measures of scarcity suffer from serious conceptual shortcomings. If attention is confined to exhaustible resources, even here, what constitutes scarcity rent (or 'user cost' or 'royalty rate') and how it changes over time remains controversial. In the standard Hotelling model, it is argued that the scarcity rent rises *monotonically* over time at the social rate of discount. In contrast, both Heal (1976) and Hanson (1980), using very different cost functions, show that the scarcity rent must decline monotonically to zero, as the resource approaches exhaustion.

On the other hand Solow and Wan (1976) analyze conditions under which resource extraction costs rise with the depletion of higher grade deposits and extraction turns to lower and lower grade ores. They argue that along the optimal path the shadow price of a resource rises at the real rate of interest, but that the difference between price and marginal extraction cost, which they call 'degradation cost,' declines monotonically over time to zero.

The divergence of views is replicated in the empirical studies too. For instance, Barnett and Morse (1963), Barnett (1979) and Johnson et al. (1980) all find that unit extraction costs in real terms have declined, so that these resources are in some sense becoming less scarce. Nordhaus (1974) also found that real prices of 11 major minerals fell over the period 1900 to 1970. In contrast, Smith (1979), Slade (1982), and Hall and Hall (1984) all find that real prices of natural resources are rising.

If we turn to renewable resources such as fish and forestry, it is not at all clear whether scarcity rent is deemed appropriate at all. The forestry literature is mainly concerned with optimal harvesting, although a number of forest ecologists have argued that forests are fast 'disappearing,' so that the remaining forests are becoming scarce (FAO, 1993). But as far as we are aware no attempt has been made to compute a shadow price or scarcity rent of forests in mitigating global warming; it is this which is a central objective of this paper.

In an attempt to reconcile the divergent and conflicting approaches to scarcity rent of natural resources, Farzin (1992) presents a generalized model which shows that *in general* the scarcity rent is non-monotonic. He shows that rising scarcity rent of exhaustible resources is a special case. He also presents an important corollary, that when the discount rate is zero (due say to intergenerational equity considerations) then contrary to the Hotelling rule, scarcity rent of an exhaustible resource always decline monotonically over time to zero, irrespective of the form of the extraction cost function.

This paper is concerned with valuing forests for an important ecological function, namely the ability of forests (at least those in Canada) to sequester carbon and so mitigate global warming. It turns out that Canadian forest are a net 'sink' of carbon–they capture more carbon than the release to the atmosphere (see Appendix 1 for details). But quite apart from being a carbon sink, forests hold soil and prevent erosion; they purify water and play an important role in the water cycle; they help maintain biodiversity; they are hosts to entire ecosystems. However, all of these other ecological functions will be ignored in this paper; the focus is entirely on forests as a carbon sink. An optimal control model (that maximizes value-added) is used to derive a shadow price of forests from the optimal solution, and the value of forest is computed via econometric techniques, with value-added as numeraire. From this the marginal social opportunity cost of forests per hectare is computed. It is argued that it is this opportunity cost that must enter any harvesting decision model, or any other social cost-benefit calculation involving the use of forests.

Our model is consistent with the results of Farzin (1992), mentioned above. The computed shadow price of forests obtained by normalizing the shadow price which respects the ecological constraint, is shown to be non-monotonic in general, and a monotonic shadow price can be obtained as a special case. However, the special case (Model 1) is instructive as an introductory thought experiment. We then use trend analysis (Model 2) and an ARIMA model (Model 3) to determine the marginal social opportunity cost of a hectare of forest.

The paper is organized as follows. In section one, the assumptions of the model are stated. In section two, the theoretical model is stated, and in section three, the model is estimated using standard econometric techniques.

SECTION ONE

Assumptions

1. Let a social planner maximize a social welfare functional V, where W is a functional of natural and human capital stock $x(t)$, a set of control variables $u(t)$, where $u(t)$ are policies such as harvesting forests, investing in conservation or other policies that improve x, and time t.
   Hence let

$$W = \frac{\max}{u(t)} \int_0^T [V(x(t), u(t), t)] dt \qquad (1)$$

   which is quasi-concave in the arguments, and V stands for value-added. The functional w embodies a Paretian judgement such that valued-added is maximized, irrespective to whom it may acrue.

2. The vector $x(t)$ is the state at time t of the natural and human made capital stock, and the time path of x, i.e., $\dot{x}$ is governed by

$$\dot{x} = f(x(t), u(t), t) \qquad (2)$$

3. Equation (2), the equation of motion, is convex. The policies $u(t)$ modify future $x(t)$. The first component of the vector $x(t)$ is $x_1(t)$ which represents 'the state of the forest.' As it is the ONLY component of the vector of interest in this paper, its subscript can be dropped without any possibility of confusion.

$$x(t_0) = x_0 \qquad (3)$$

4. The following initial and terminal conditions are given:

   where x_T is defined to be sustainable (Dore and Ward, 1995), beyond T to infinity.

$$x(T) = x_T \qquad (4)$$

5. The social value of value-added is "valued at par" only when there is no externality, i.e., when carbon emissions are just equal to the carbon sequestration by forests. Furthermore, the social value of value-added declines proportionally with the excess of carbon emissions over carbon sequestration.

SECTION TWO: THE MODEL

Logical Implications of the Assumptions

The logical implication of assumptions 1 to 3 is that a dynamic optimal solution to equations 1, 2, and 3 exists, where the optimized value of W is W^*. This in turn implies that there exist costate variables $\gamma(t)$, which are positive numbers.

From the maximum principle, it follows that:[1]

$$\frac{\partial W^*}{\partial x^*} = \gamma^*(t), \quad \forall\, t_0 \le t < T \qquad (5)$$

As usually interpreted, the costate variable (to a convex problem) measures the sensitivity of the optimized value of the objective function to a slight relaxation in the constraint.

Define the reciprocal of $\gamma(t)$ to be $\lambda(t)$. It follows that the latter are also positive numbers. Writing equation 5 in discrete time, we have:

$$\frac{\Delta W}{\Delta x} = \gamma(t) = \frac{1}{\lambda(t)} \qquad (6)$$

or,

$$\Delta x = \lambda(t)\Delta W \tag{7}$$

From assumption 1, ΔW is value-added in each time period, and $\lambda(t)$ is the shadow price of forests.

It follows that as the right hand side of equation (7) has the dimension of value (the 'value' of value-added weighted by the shadow price), the left hand side must also have a value dimension: the change in the 'state' of forests is the *flow value of the forests,* with value-added as numeraire. That is, we are valuing forest services in terms of a conventional macroeconomic category, namely, value-added.[2] It is equation (7) that is estimated using econometric techniques.

Modelling Carbon Sequestration

It has been established by Kurz, Apps, et al. (1992) that Canadian forests are a net sink for carbon because of its age composition and species distribution.

Let q(t) be carbon sequestered per hectare in MT. Then q(t) is given by:

$$q(t) = \int_1^\kappa \int_0^t b \cdot \sigma(\tau)d\tau \cdot d\kappa - \int_0^\kappa \int_0^t z \cdot \rho(\tau)d\tau \cdot d\kappa \tag{8}$$

where, b   = number of trees of each species per hectare
κ = number of species of trees
$\sigma(\tau)$ = the age distribution of trees of each species per hectare
z   = number of trees of each species per hectare lost due to fire, insect damage or other human activities
$\rho(\tau)$ = probability of loss

Therefore q(t) is the net carbon sequestered per hectare in time period *t*. Let the total area of forest at time *t* be *a(t)*. Then total carbon sequestered (in MT) is

$$Q(t) = q(t) \cdot a(t) \tag{9}$$

This completes the formal modelling of carbon sequestration, or 'uptake.' Details of the estimation of carbon sequestered are given in Appendix 1.

Modelling the Shadow Price

The reasons why the concept of scarcity rent is regarded controversial by Brown and Field (1978) is that first, it is non-observable, and there is no standard method of numerical estimation. Second, the most appealing version of a shadow price is the costate variable that would be derived from dynamic optimization, as done in this paper. As stated earlier the costate variable is the marginal increase in value-added due to a slight relaxation in the constraint; and its reciprocal is a shadow price of forests in terms of value-added. But before a costate variable can be used empirically, strictly speaking, the dynamic optimization model of equations 1 to 5 must by solved numerically. This raises the serious issues of observability and controllability (Dore, 1995), which cast doubt as to whether the model can in fact be solved numerically.

The failure of observability and controllability occurs because the dynamic constraint given in equation 2 is not based on some invariant physical law, such as the laws of physics and engineering. Thus while observability and controllability hold for firing rockets into space, they fail for most large scale economic applications.

Third, even if it were possible to solve for the numeral values of the costate variables and its reciprocal $\lambda(t)$, the complementary slackness conditions mean that the costate variable is positive if and only if the constraint is fully binding. For our problem it means that the "forests must be fully utilized" with no excess capacity. Thus the costate variable measures the sensitivity of the objective function due to a slight relaxation of the constraint only when the constraint is binding, but is otherwise zero. As noted by Brody (1970), in a different context, this does not yield a satisfactory index of relative scarcity.

The shadow price of forests *should be* an index of the scarcity of forests. Now forests absorb carbon emissions which are generated in the production of value-added, using fossil fuels. In fact forests may be seen as joint social capital required in the production of output and value-added. This social capital has 'excess' capacity if emissions of carbon are less than the carbon sequestered by forests per year. The carbon sequestered is called *uptake*. When uptake exceeds emissions there is excess capacity, and some forested areas can be cut down for alternative uses. However, when emissions exceed uptake, forests become 'scarce' and the index of scarcity must be greater than one.

From the Maximum Principle, it follows that:

$$\gamma(t) \leq 0 \quad \forall t_0 \leq t < T \tag{10}$$

from which it follows that:

$$\dot{\lambda}(t) \geq 0, \quad \forall \, t_0 \leq t < T \tag{11}$$

If the objective function is strictly concave, then the costate variable is *monotonic decreasing*.

Consequently,

$$\dot{\lambda}(t) > 0 \quad \forall \, t_0 \leq t < T \tag{12}$$

Hence $\lambda(t)$ is the correct index of scarcity, for it increases when the stock of forests decreases. If the objective function is strictly concave, then the index of scarcity is *monotonic increasing*. This is of course a special case, as argued by Farzin (1991).

In order to obtain a cardinal index of scarcity, we make assumption 5. According to this assumption, the social value of value-added (or output) is "valued at par" only when there is no externality, i.e., when carbon emissions are just equal to the carbon sequestration by forests. Furthermore, if the value of value-added declines proportionally with the excess of carbon emissions over carbon sequestration, then the *magnitude* of the externality is taken into account. This suggests the normalization of the shadow price to 1 when emissions equal uptake.

It follows that the shadow price is of the form:

$$\lambda(t) = \lambda \, (e(t), \, Q(t)), \tag{13}$$

where $e(t)$ are emissions of carbon in MT and $Q(t)$ is uptake of carbon in MT.

The above form of the equation yields a cardinal index of relative scarcity which over time need not necessarily be monotonic; monotonicity will hold only as a special case. For example, monotonicity is obtained if $e(t)$ is growing but $Q(t)$ is held constant, as in Model 1 in section three.

Our econometric estimates in section three, use equation (13) to compute and estimate a cardinal measure of relativity scarcity over time. The interpretation is straightforward: when value-added generates emissions greater than uptake, $\lambda(t) > 1$, i.e., forests are valued at a premium, and when emissions are less than uptake, $\lambda(t) < 1$, and forests are valued at a discount, for in that latter case we can produce more emissions by producing more value-added without adding carbon emissions into the 'Canadian' atmosphere.

As stated above, in general $\lambda(t)$ will be non-monotonic and that monotonicity can be obtained as a special case. It will be shown that this special case occurs when the uptake $Q(t)$ is held constant, and when the analysis is confined to trend analysis only. Such a 'ceteris paribus' thought experiment is instructive as a first approximation in the valuation of forests (Model 1) below. When $Q(t)$ is not held constant, we get a different (non-monotonic) time path of the shadow price $\lambda(t)$. In Section 3, two such models are considered depending on the method of estimation, where Model 2 is a straight linear regression over time; Model 3 is an ARIMA model.

This completes the presentation of the model. The next section is an econometric estimation of the main equation, which is equation 7, reproduced below:

$$\Delta x(t) = \lambda(t)\Delta W(t) \tag{7}$$

SECTION THREE:
ECONOMETRIC ESTIMATES

Before presenting the estimations, it might be useful to consider Figure 1, which is a graph of Canadian Carbon emissions in MT, as well as the sequestration of carbon by Canadian forests, for the period 1961 to 1991. A time series of atmospheric carbon emissions for Canada over the period 1961-1991 is provided by Marland, Andres and Boden (1994) and published by CDIAC. These authors report a continuing growth in emissions from 1961 until 1980. There was a drop in emissions after 1980, and the pattern from 1980 to 1987 was erratic but with a small downward trend, as shown below. During the last two years of the series, emissions have decreased 7% after peaking in 1989 at a level of 120.7 MT C. It is clear that Canadian emissions of Carbon have been increasing, although the rate of growth has become volatile, with a slight downward trend from 1980 onwards. This slowdown is in part a legacy of the oil shocks of 1973 and 1979, and in part due to structural changes within the economy, with the relative decline of manufacturing in most industrialized nations. From about 1969, Canadian emissions began to exceed the amount of carbon sequestered by Canadian forests.

Data on carbon sequestration (or uptake) is only just beginning to be refined, a subject that is itself quite complex. Details on the data on carbon sequestration as considered in Appendix 1.

FIGURE 1. Canadian Carbon Emissions versus Carbon Uptake by Canadian Forests

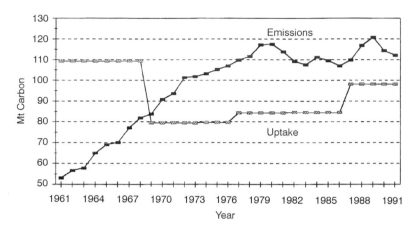

We now proceed with the presentation of our empirical results. As stated above we report three model estimates.

Model 1: Trend Analysis

Model 1 is a first approximation, obtained by fitting a time trends to emissions holding uptake constant, a time trend can be fitted to Canadian manufacturing value added for the period 1961-1991.

Emissions

A linear regression of carbon emissions on time is fitted to annual Canadian data. In light of the comments made by Marland et al. (1994), we are interested in testing for a structural break in 1980. The possibility of such a break is intuitively appealing, given the oil price shock that occurred in 1979 and the structural changes in manufacturing considered to have taken place following the 1980-1982 recession. The test employed is based on a dummy variable technique.

The basic model is

$$\text{Emissions}_t = a + bt + \mu_t$$

We define the dummy variable as follows:

$D_1 = 1$ for the period 1980 onward
$D_1 = 0$ otherwise

To test whether the structures for the two periods (prior to 1980, and the period after 1980) are different the models assumes the following specification:

$$\alpha = \alpha_0 + \alpha_1 D_1 \qquad \beta = \beta_0 + \beta_1 D_1$$

Thus, the unrestricted model is:

$$\text{Emissions}_t = \alpha_0 + \alpha_1 D_1 + \beta t + \beta_1 (D_1 t)$$

The estimated models are:

Prior to 1980: $\text{Emissions}_t = \alpha + \beta_0 t$

1980 onward: $\text{Emissions}_t = \alpha_0 + \alpha_1 + (\beta_0 + \beta_1) t$

Thus, a failure to reject the null hypothesis that $a_1 = b_1 = 0$ would indicate the presence of a structural break at 1980.

A Lagrange Multiplier test on the residuals following a first run of the model indicated the presence of first order serially correlated errors (p-value = 0.0080). Based on White's test for heteroscedasticity, we reject the null hypothesis that the errors have non-constant variance (p-value = 0.1218). The regression was estimated again, with adjustments to the standard errors of the estimated coefficients (based on the estimation of a consistent covariance matrix).

The fitted model is:

Prior to 1980: $\text{Emissions}_t = 50.76 + 3.63\,t$
$$(0.17)\ \ (0.43)$$
$$t = (34.35)\,(21.91)$$
$$R^2 = 0.9988$$
$$\text{d.f.} = 27$$

1980 onward: $\text{Emissions}_t = 54.60 + 55.56 + (3.63 - 3.39)t$
$$(11.11) \qquad (0.44)$$
$$t = (4.32) \qquad (-4.69)$$

RMSE = 3.60

i.e., $\text{Emissions}_t = 110.16 + 0.24t$

A joint significance test of the null hypothesis that $\alpha_1 = \beta_1 = 0$ lead to rejection of the null (p-value = 0), and we conclude that there was a structural break in 1980.

Value Added

Statistics Canada provides us with a time series of value-added in Canadian Manufacturing. Value-added has essentially continued to grow over time; however, beginning in about 1980 levels of value-added have become more variable. Of course, the period 1980-1991 includes two major disturbances, the deep recession of 1981-1982 and the onset of the lesser recession that began in 1990.

A linear time trend is fitted to annual data on value-added in Canadian Manufacturing.

The model is simply:

$$\text{Value Added}_t = \alpha + \beta t + \mu_t$$

As in the case of carbon emissions, the results of a first run of the model indicated the presence of first order serially correlated errors (p-value = 0.0088). White's test indicates the presence of heteroscedastic errors (p-value = 0.0060). The model was estimated again, with adjustments made to the standard errors as above.

The fitted model is:

$$\text{Value Added}_t = \begin{array}{cc} 46.90 & + \; 2.01t \\ (1.97) & (0.17) \end{array}$$
$$t = (23.74) \; (11.48)$$
$$R^2 = 0.9954$$

RMSE = 5.67   d.f. = 29

The Value of the Forests

The estimated shadow price is shown in Figure 2.[3] Note that, except for the structural break, a monotonic shadow price is obtained as (a) uptake is held constant at 89 Mt Carbon, and (b) a linear trend is fitted. While this is a special case, it is nevertheless instructive mainly as a thought experiment. The estimated value of Canadian forests is then given in Figure 3, with the dotted lines giving the 95 percent

FIGURE 2. Estimated Shadow Price

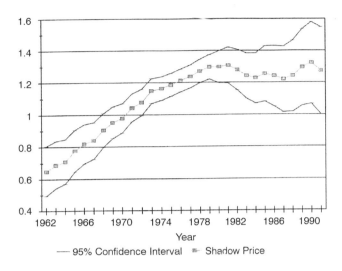

Year

— 95% Confidence Interval ▣ Shadow Price

FIGURE 3. Estimated Value of Forests for Carbon Sequestration, Canada, 1962-1991

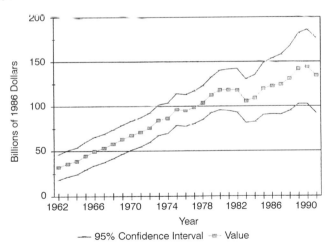

Year

— 95% Confidence Interval ▣ Value

confidence bands based on the calculated confidence interval for the individual in-sample point forecasts, or individual predictors. For a comparative perspective, consider Figure 4, where the estimated value of the forests is expressed as a percentage of GDP. Note that the 1990 figure of the value of the forests as a proportion of real GDP is close to

23 percent. But the important estimate is the value of forests per hectare: this is the *marginal social opportunity cost of one hectare of forest*. The value of the forests per hectare is shown in Figure 5. The 95 percent confidence intervals yield the "high" and the "low" estimates.

FIGURE 4. Estimated Value of Forests for Carbon Sequestration as a Percentage of GDP

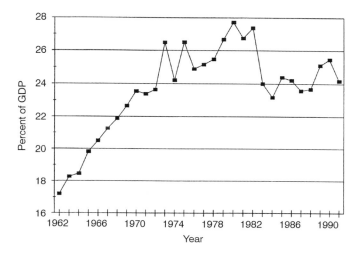

FIGURE 5. Forest Value per Hectare, Canada, 1962-1991

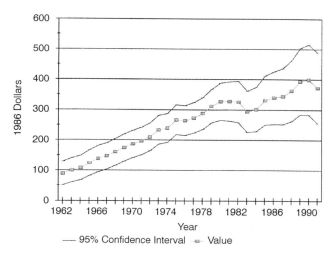

Model 2: Trend Analysis with Variable Uptake

Next, we relax the assumption of a constant uptake. Here we also exploit the information in the residuals and estimate auto-correlation models, using Cochrane-Orcutt as the method for estimation.

The fitted model is:

Prior to 1980:   Emissions$_t$ = 54.6 + 3.26 t
                                 (5.94) (0.43)
                       t = (9.19) (7.63)
                                           R^2 = 0.9992
                                           d.f. = 25

1980 onward: Emissions$_t$ = 54.60 + 66.38 + (3.26-3.56)t
                                     (15.36)   (0.76)
                            t = (4.32)   (-4.69)

RMSE = 3.05

i.e., Emissions$_t$ = 120.98 − 0.30t

A Dickey-Fuller unit root test on the residuals returned a test statistic value of −24.6159, leading to the rejection of a unit root. A joint significance test of the null hypothesis that $a_1 = b_1 = 0$ lead to rejection of the null (p-value = 0.0003), and we conclude that there was a structural break in 1980. In fact, contrary to the interpretation given by Marland et al. (1994), these results show that there is a slight downward trend in Canadian carbon emissions since 1980.

In the case of value added, the fitted model is:

Value Added$_t$ = 50.59 + 1.78t
                       (5.1)   (0.21)
             t = (9.91) (6.91)
                                 R^2 = 0.9965
RMSE = 5.11   d.f. = 27

White's test failed to reject the null of homoscedastic errors at the 5% significance level but not at the 1% significance level. At the small risk of failing to reject a false null, the residuals are considered to have constant variance. A Dickey-Fuller unit root test run on the residuals returned a test statistic of −22.3342. This test rejected a unit root.

It will be seen from Figure 6 that the shadow price is no longer completely monotonic. Figure 7 then gives the corresponding value of the forests, together with confidence intervals. Figure 8 presents the value of the forests as a proportion of real GDP, and Figure 9 gives the value, in constant 1986 dollars, of a hectare of forest. For instance, the High and Low estimates for 1986 and 1990 are given in Table 1.

FIGURE 6. Estimated Shadow Price

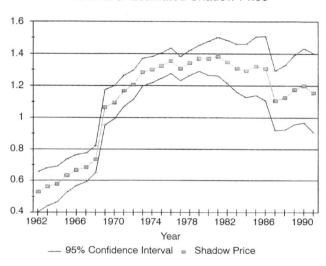

FIGURE 7. Estimated Value of Forests for Carbon Sequestration, Canada, 1962-1991

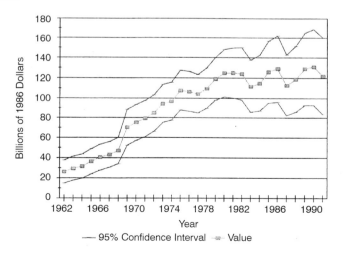

FIGURE 8. Estimated Value of Forests for Carbon Sequestration as a Percentage of GDP

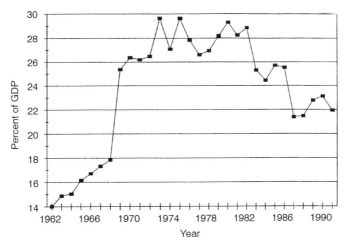

FIGURE 9. Forest Value per Hectare, Canada, 1962-1991

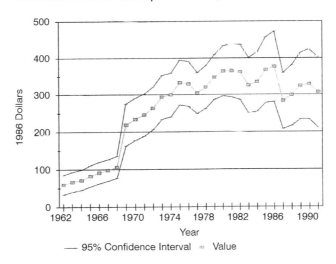

Model 3: ARIMA Model Estimates

An alternative approach involves fitting a model based on the notion that present levels of emissions and value-added are related to past levels. Since the time series are so short, we began by first fitting pure auto-regressive models of order n.

TABLE 1. Value of Forests per Hectares in 1986 Constant Dollar

	Low	Estimate	High
1986	259.9	347.0	434.4
1990	266.2	378.3	490.4

Source: Figure 9

Carbon Emissions

In the case of carbon emissions, a first order auto-regressive model was fitted to the data. An LM test on the residuals indicated the presence of first order serially correlated errors (p-value = 0.0014). Consequently, the series was first differenced to make it stationary. Three alternative unit root tests were carried out (Dickey-Fuller, Phillips-Perron, and Stock-Watson) on this series. Each test rejected a unit root. For full details on the justification of the ARIMA modelling in this paper, see details of the unit root tests given in Appendix 2.

An ARIMA (1, 1, 0) Model was then fitted to annual data on Canadian Carbon Emissions. (Higher order auto-regressive terms were found to be statistically insignificant.)

The fitted model is:

$$E'_t = 0.4606 \, E'_{t-1}$$

where $E'_t = (1 - B) E_t$

or $(1 - 0.4606B)(1 - B) E_t = a_t$ where a_t is the white noise errorterm

    (0.1660)
    t = (2.7744)
df = 28
RMSE = 3.7192

Value Added

In the case of value-added, a first order auto-regressive model was also fitted to the data. An LM test on the residuals indicated the presence of first order serially correlated errors (p-value = 0.0304), and therefore, this series too was first differenced to make it stationary. Three alternative unit root tests were carried out (Dickey-Fuller, Phillips-Perron, and Stock-Watson) on this series. Each test rejected a unit root, at the 5 percent level. (See Appendix 2 for details.) The model based on first differences was then estimated. Again additional autoregressive terms were found to be statistically insignificant; however, the coefficient on the dependent variable was found to be statistically insignificant. For this reason the series was second differenced, with unit root tests rejecting the presence of a unit root as above. An ARIMA (1, 2, 0) Model was then fitted to annual data on Value-Added in Canadian Manufacturing.

$$VA'_t = -0.4254 \, VA'_{t-1}$$

where $VA'_t = (1 - B)^2 \, VA_t$

or $(1 + 0.4254B)(1 - B)^2 \, VA_t = a_t$ where a_t is the white noise error term
$\quad$ (0.1756)
$t = (-2.4224)$
$df = 27$
$RMSE = 6.3613$

Value of the Forests

Figure 10 gives the estimated shadow price with the 95 percent confidence intervals. Figure 11 gives value of the forests, again with the confidence intervals. Figure 12 gives the value of the forests as a proportion of GDP, and Figure 13 gives the value of the forests per hectare in 1986 constant dollars. The High and Low estimates of 1986 and 1990 are given in Table 2. Finally Table 3 summarizes the most important results of the three models.

Finally, we compare the results of Models 2 and 3. Note first that the value of forests per hectare falls between 1986 and 1990, because the forested area goes up. The 1990 ARIMA estimate (Model 3) is higher because the in-sample forecast for emissions is higher, which is in turn higher due to a higher emissions level in the previous year (1989). We would expect the ARIMA model to pick this up. However, from Figure 14 it is clear that the two models yield fairly close estimates. This is the case even when there was a structural break incorporated in Model 2 but none in Model 3, because the ARIMA estimation procedure reflects the structural changes endogenously: the fitted in-sample estimate of 1981 shows a sharp 'drop.'[4] The fall shows up in

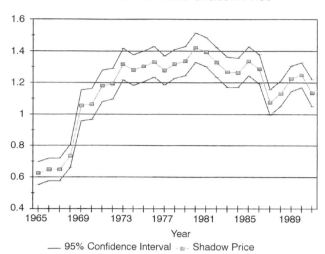

FIGURE 10. Estimated Shadow Price

FIGURE 11. Estimated Value of Forests for Carbon Sequestration, Canada, 1965-1991

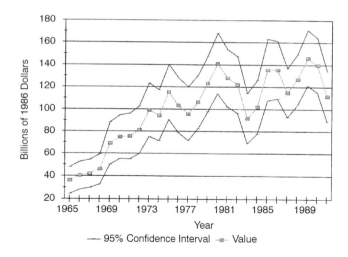

1981 because the ARIMA model is of course (an auto-regressive) distributed lag model.

The ARIMA model may indeed be a better forecasting tool, for if we were to incorporate the marginal social opportunity cost of a hect-are of forest in any cost benefit analysis, or into an optimal harvesting

FIGURE 12. Estimated Value of Forests for Carbon Sequestration as a Percentage of GDP

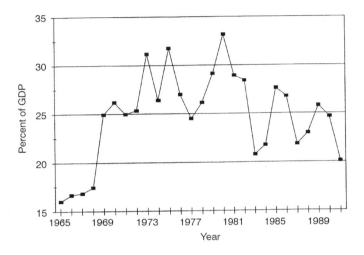

FIGURE 13. Forest Value per Hectare, Canada, 1965-1991

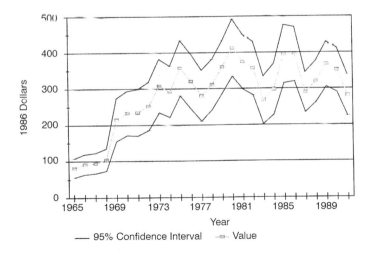

model, the most appropriate thing to do would be to use the ARIMA estimates to produce ex ante forecasts, say for 1993 or even 1995; that is, for time periods beyond the sampling period. For these two reasons alone, we prefer the ARIMA estimates, as they are likely to be more useful.

TABLE 2. Value of Forests per Hectare in 1986 Constant Dollar

	Low	Estimate	High
1986	319.4	394.7	470.0
1990	289.8	351.1	412.4

Source: Figure 13

TABLE 3. Selective Summary of the Estimates: Shadow Price and the Value of the Forests per Hectare

	Shadow Price			Value of the Forests		
	Low	Estimate	High	Low	Estimate	High
Model 1: **Trend Analysis**						
1986	1.08	1.26	1.45	259.9	347.0	434.1
1990	1.02	1.27	1.53	266.2	378.3	490.4
Model 2: **Trend Analysis**						
1986	1.10	1.31	1.51	280.5	376.5	472.5
1990	0.97	1.20	1.43	233.6	324.2	414.7
Model 3: **ARIMA Model**						
1986	1.19	1.29	1.38	319.4	394.7	470.0
1990	1.17	1.25	1.33	289.8	351.1	412.4

FIGURE 14. Value of Canadian Forests for Carbon Sequestration 1965-1991

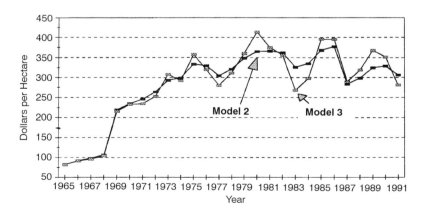

CONCLUSIONS

This paper is an attempt to estimate the social value of forests that sequester carbon in the biomass of the forests, carbon that is a byproduct of the burning of fossil fuels in the production of industrial output. It avoids the lack of invariance associated with willingness to pay (WTP) and willingness to accept (WTA) as follows. At a social optimum, WTP must also reflect marginal social opportunity costs (MSOC), so that at the optimum, MSOC = WTP. We set up a model in which: (a) the equation to be estimated as well as the shadow price is in principle generated from a dynamic optimizing framework. This is our equation (7). This equation gives the MSOC value of the forests. As the objective functional in the dynamic optimization model is the maximization of value-added, with no distributional considerations, the implied *numeraire* in the MSOC value of the forests is indeed value-added.

While the shadow price, which is the costate variable, is in principle generated by the dynamic optimization, it is subject to the restrictive complementary slackness condition, which in theory yields a zero shadow price whenever a constraint in not fully binding. The result is that the costate variable is not a complete cardinal index of scarcity. As we are using the reciprocal of the costate variable as the shadow price of the forests, the same limitation applies to the shadow price of the forests. To overcome this problem, we assume that value-added is valued at par only when there is no externality, i.e., when carbon emissions are equal to carbon sequestered by the forests. We thus make the shadow price a function of the ratio of carbon emissions to sequestration. When emissions are just equal to sequestration, at that time, the value of the forests is just equal to the value added made possible by forests as social (overhead) capital. Consequently at that time value added is valued at par. However, as soon as carbon emissions exceed carbon sequestration, the marginal social value of value-added is less than its constant dollar value, because of the externality. Reciprocally the value of the forests is higher than the constant dollar value of value-added. In this formulation the shadow price is a cardinal index of the relative scarcity of the forests, which are really a joint input in the production process that uses carbon-emitting fossil fuels.

The econometric estimates of the shadow price of the forests and the resulting MSOC value of forests per hectare are obtained by a time trend model (Model 2) and an ARIMA model (Model 3). We regard the ARIMA model estimates to more representative of current conditions, as the time trend model (Model 2) merely extrapolates from an established trend, even when structural breaks are taken into account. The estimated shadow price of the forests (in 1990) is a premium of the order of 20 percent. The corresponding MSOC value of a hectare of forest is between $290 to $412 in 1986 constant dollars. Naturally the estimates are subject to the limitations of the data.

NOTES

1. Actually, along the optimal solution,

$$\frac{\partial w^*}{\partial x^*} = y^*(t)$$

but we are only interested in the first component of the vector x. The equation given above is a standard property of the maximum principle. Chiang (1992, pp. 206-207) states it without proof, but Léonard (1987) gives an explicit proof of this property.

2. In the absence of indirect taxes, the sum of value added is nothing but GDP.

3. In all the figures in this paper, the dashed lines give the 95 percent confidence intervals.

4. Strictly speaking a structural break can be introduced if the sample were split into two: pre-1980 sample and a post-1980 sample. But this would give a sample of 10 observations, requiring the estimation of 4 parameters. This would drastically reduce the degrees of freedom. In any case, as argued in the text, we do not believe that a structural shift is in any way biasing our ARIMA results. To test the robustness of our results, we estimated an alternative model (Model 4) which was ARIMA (0,1,1) which showed no structural break and was statistically significant. However, the numerical results were virtually identical with that of Model 3. For this reason, the results of Model 4 are not reported in this paper. But they do confirm the robustness of our numerical estimates.

REFERENCES

Barnett, H. J. (1979). *Scarcity and Growth Revisited*. In Smith (1979). Scarcity and Growth Revisited. Baltimore: Johns Hopkins University Press (for Resources for the Future), pp. 163-217.

Barnett, H. J. and Morse, C. (1963). *Scarcity and Growth: The Economics of Natural Resource Availability*. Baltimore: Johns Hopkins University Press (for Resources for the Future).

Brody, A. (1970). *Proportions, Prices and Planning*, Amsterdam: North Holland.

Brown, G. M., Jr. and Field, B. C. (1978). Implications of alternative measures of natural resource scarcity. *Journal of Political Economy*, 86(2): 229-43.

Chiang, A. C. (1992). *Elements of Dynamic Optimization*. New York: McGraw-Hill.

Dore, M. H. I. (1995). Dynamic games in macro models: a critical appraisal. *Journal of Post-Keynesian Economics*, 18(1): 107-123.

Dore, M. H. I., and Ward, A. J. (1995). *Modelling Sustainable Development: A Nonlinear Analysis*. Brock University, mimeo.

Farzin, Y. H. (1992). The time path of scarcity rent in the theory of exhaustible resources. *The Economic Journal*, 102: 813-830.

Fisher, A. C. (1979). *Measures of Natural Resource Scarcity*. In Smith (1979b), pp. 249-75.

Farzin, Y. H. (1984). The effect of discount rate on depletion of exhaustible resources. *Economic Journal*, 92: 841-51.

Hall, D. C., and Hall, J. V. (1984). Concepts and measures of natural resource scarcity with a summary of recent trends. *Journal of Environmental Economics and Management*, II: 363-379.

Hanson, D. A. (1980). Increasing extraction costs and resources prices: some further results. *Bell Journal of Economics*, 11: 335-42.

Heal, G. M. (1976). The relationship between price and extraction cost for a resource with a backstop technology. *Bell Journal of Economics*, 7: 371-8.

Horbulyk, T. M. and Wm. A. Ross. (1993). Potential to sequester carbon in Canadian forests: a comment. *Canadian Public Policy*, XIX:4:463-468.

Johnston, M. H., Bell, F. W. and Bennett, J. T. (1980). Natural resource scarcity: empirical evidence and public policy. *Journal of Environmental Economics and Management*, 7:256-71.

Léonard, D. (1987). Co-state variables correctly value stocks at each instant: a proof. *Journal of Economic Dynamics and Control*, 11: 117-122.

Kurz, W. A., Apps, M. J., et al. (1992). *The Carbon Budget of the Canadian Forest Sector: Phase 1*. ENFOR Information Report NOR-X-326 (Edmonton: Forestry Canada, Northwest Region).

Marland, G., Andres, R. J. and Boden, T. A. (1994). Global, regional and national CO_2 emissions. pp. 505-584. In T. A. Boden, D. P. Kaiser, R. J. Sepanski, and F. W. Stoss (eds). *Trends '93: A Compendium of Data on Global Climate Change*. ORNL/CDIAC-65. Carbon Dioxide Information Analysis Center, Oak Ridge National Laboratory, Oak Ridge TN, USA.

Nordhaus, W. D. (1974). Resources as a constraint on growth. *American Economic Review*, 64: 22-26.

Slade, M. E. (1982). Trends in natural-resource commodity prices: and analysis of the time domain. *Journal of Environmental Economics and Management*, 9: 122-37.

Smith, V. K. (1979). *Scarcity and Growth Reconsidered*. Baltimore: Johns Hopkins University Press (for Resources for the Future).

Solow, R. M. and Wan, F. Y. (1976). Extraction costs in the theory of exhaustible resources. *Bell Journal of Economics*, 7: 359-70.

van Kooten, G. C., Arthur, Louise M., and Wilson, W. R. (1992). Potential to sequester carbon in Canadian forests: some economic considerations. *Canadian Public Policy*, XVIII:2:127-38.

van Kooten, G. C., Thompson, W. A., and Vertinsky, I. (1993). Economics of reforestation in British Columbia when benefits of CO_2 reduction are taken into account. In W. Adamowicz, W. White, and W. E. Phillips (eds), *Forestry and the Environment: Economic Perspectives*, Wallingford, UK: Commonwealth Abstract Bureau International.

APPENDIX 1

Determination of the Amount of Carbon Sequestered at Time t.

Essentially, the net flux of carbon (C) between the forest ecosystem and the atmosphere determines whether forests are part of the problem or part of the solution with regard to changes in the concentration of CO_2 in the atmosphere. We are interested in obtaining the most appropriate estimate of carbon uptake by forest biomass, reported in the scientific literature. In their examination of the potential role of reforestation policies in offsetting Canada's emissions of atmospheric carbon dioxide, van Kooten, Arthur and Wilson (1992, pp. 127-38) determine the rate at which carbon is sequestered on Canada's forested lands by multiplying forest land area (ha) in each ecoclimatic province by the annual productivity of the forest in each zone (m^3/ha/yr) and by the average amount of carbon in green timber (260 kg/m^3). The results of this calculation are 203 million tons (MT) per year C sequestered by Canadian forests. However, there are problems with using an average carbon content value in that it is too general. Consequently, van Kooten, Thompson, and Vertinsky (1992 cited in Horbulyk and Ross, 1993, pp. 463-7) employ a different set of assumptions to estimate a lower rate of carbon sequestration on the same forest lands. The new value is 157.1 MT per year, representing a decrease of 23% relative to that arrived at in van Kooten et al. above. This decrease is due to the use of specific carbon content of wood estimates for the mix of tree species found in each ecoclimatic province. This change in approach effectively reduces the average rate of carbon sequestration from 260 kg/m^3 to 200 kg/m^3. However, this figure (157.1 MT per year) is probably an overestimate. Although a key criticism made by Horbulyk and Ross of the figure obtained by van Kooten et al. has been taken into account in the revised estimate, Horbulyk and Ross also criticize van Kooten et al. for their use of mean annual increment data for average forest productivity as if they were net values from which deductions had been made for all mortality losses due to fire and pest epidemics. Horbulyk and Ross make the important point that where losses occur, forests are not only unable to sequester carbon at average rates but also release atmospheric carbon. New estimates for Canada of the 1990 carbon uptake by forest biomass and forest soils are revealed in *The Carbon Budget of the Canadian Forest Sector: Phase I* (Kurz, Apps, et al., 1992). Given the scope of the analysis, we believe that their estimates are the best available to date. This data is used to approximate our equations 8 and 9.

The results of the standard simulation run, show that prior to accounting for disturbances, the biomass C pools of Canadian forest ecosystems sequestered an estimated 109.3 MT of C, comprising the 92.0 MT increase in the biomass C pool and the 17.3 MT C transferred to the soil C pools as detrital material (litter fall and tree mortality). Carbon uptake is distributed across all forests in Canada (Kurz, Apps, et al., p. 41-44).

Every year a small proportion of the Canada forests is disturbed by fires, insects or harvesting. In the reference year 1986, disturbances released 20.3 Mt of the C sequestered in forest biomass directly to the atmosphere. Disturbances also transferred biomass C to the forest products sector (44.2 MT C harvested biomass) and to the soils pools (55.4 MT C). The latter transfer consisted of slash and dead standing trees after harvesting, fires and insect-induced tree mortality (included in the soil C pools in this model).

Thus, we conclude that the biomass carbon pools of the Canadian forest ecosystem sequestered approximately $109.3 - 20.3 = 89$ MT of Carbon. C released through disturbance but transferred to the forest products sector or soils pools is not deducted, as it effectively remains sequestered. The authors of the *Carbon Budget* caution that this is a single time step simulation and that extrapolation would be inappropriate. However, to facilitate the analysis of carbon value several assumptions are made that allow us to create a time series of carbon uptake by Canadian forests, however imperfect. We assume that species composition and age distribution are fairly constant over the period 1961-1991. We denote uptake at time t as $Q_t = q\,\alpha_t$, where q = rate of carbon sequestration = 89 MT C as calculated above and α_t = forest area at time t. Apps and Kurz identify 1986 as the reference year for the model simulation, utilizing data from the period 1981-1989. For example, estimates of carbon released due to wildfire are based on the 10 year average of area burned. Data on the extent of the forest (a_t) is sketchy since forest inventories are done only intermittently. The period 1961-1991 saw the extent of the forests drop from about 443 million hectares to 322 million ha through the 1970s and return to about 400 million ha toward the end of the 1980s. Based on this data, the 10 year average (1981-1989) of forest area is about 361 million hectares. Thus, we arrive at a value for q of $89/361 = 0.2465$ MT C per million hectares or about 1/4 of a mega ton for every million hectares of forest. This shows how we obtained a reasonable approximation for our equations 8 and 9, given the limited availability of the data.

APPENDIX 2

UNIT ROOT TESTS

Emissions

In general, econometric methods, including ARIMA modelling, are valid only if the time series being modelled is stationary. We took first differences of emissions based on the fact that the original time series of emissions was non-stationary. We determined this through a casual inspection of the correlogram, à la Box-Jenkins. However, we can also use the Augmented Dicky-Fuller (ADF) tests: the T (rho-1) test and the regression t test. The T (rho-1) test statistic on the original emissions series turns out to be -3.72, while the critical value at 5% is -21.7. As $-3.72 > -21.7$, we accept the null that there is a unit root. In terms of the regression t test, the statistic returned is -1.31, and the critical value is -3.41. Thus we again accept the null that there is a unit root. Therefore, the results indicate that the series must be differenced. The next question is the determination of the number of auto-regressive terms that will be necessary. We tried including additional lagged terms, but this proved to be not significant. We conclude that one lagged term is all that is required.

The T (rho-1) test statistic returned on the (first) differenced series is -24.32, whereas the critical value at 5% is -21.7. Therefore we reject the null and conclude that there is no unit root in the first differenced series.

Similarly, in terms of the t test, the test statistic is -4.25 but the critical value is -3.41 at 5%. Again we reject the null. Thus, we have shown that the first differenced series is stationary and we do not have any of the problems of an explosive error term as described by Ramanathan (1992).

Value Added Series

When ADF is run on the original series, the T (rho-1) test statistic is -15.86, whereas the critical value is -21.7 at 5%. In terms of the t test, the t value is -3.03, whereas the critical value is -3.41. Thus we accept the null of a unit root and conclude that the series must be differenced. Again, the question is the determination of the number of auto-regressive terms that will be necessary. We tried including additional lagged terms, but this proved to be not significant. We conclude that one lagged term is all that is required.

The T (rh0-1) test statistic returned on the (first-) differenced series is -23.66, whereas the critical value is -21.7 at 5%. We would conclude that there is no unit root. However, in the case of the t test, the test statistic is -2.72, whereas the critical value is -3.41. Based on this test, we would expect that there is a unit root. We therefore took second differences. When

ADF is run on the second differenced series, the t test statistic returned is −4.62 (whereas the critical value is −3.41). We conclude that second differencing is required to yield a stationary series. As in the case above, we do not have the problem described by Ramanathan of an explosive error term. However, in this case, second differencing was required because first differencing did not yield a significant t statistic on the independent variable, i.e., value-added in the last period. This Appendix justifies the differencing of the two series, based on unit root tests.

Sustainability
for British Columbia Forestry

Tony Ward

SUMMARY. Achieving forest sustainability in a low-population density society such as British Columbia (B.C.) poses different problems to those of South America. British Columbia has experienced the same problems as most of North America, in that for a century or two its forests have been plundered to support rapid economic growth. The Province has now to adjust to lower rates of extraction of forest products. Increased awareness of the services a forest provides, besides the provision of wood products, is leading legislators to revise substantially forest management practices. That ownership of Canadian forests is still almost entirely in the hands of the public sector helps the process of change. *[Article copies available for a fee from The Haworth Document Delivery Service: 1-800-342-9678. E-mail address: <getinfo@haworthpressinc. com> Website: <http://www.haworthpressinc.com>]*

KEYWORDS. Sustainability, British Columbia forestry, economic growth, forest mangement

INTRODUCTION

Under the 1867 Constitution Act, Canada's provinces hold the ownership of most forest lands within their borders, as well as mines and

---

Tony Ward is Assistant Professor, Department of Economics, Brock University, St. Catharines, Ontario, Canada L2S 3A1 (E-mail: award@spartan.ac.brocku.ca).

[Haworth co-indexing entry note]: "Sustainability for British Columbia Forestry." Ward, Tony. Co-published simultaneously in *Journal of Sustainable Forestry* (Food Products Press, an imprint of The Haworth Press, Inc.) Vol. 12, No. 1/2, 2001, pp. 153-169; and: *Climate Change and Forest Management in the Western Hemisphere* (ed: Mohammed H. I. Dore) Food Products Press, an imprint of The Haworth Press, Inc., 2001, pp. 153-169. Single or multiple copies of this article are available for a fee from The Haworth Document Delivery Service [1-800-342-9678, 9:00 a.m. - 5:00 p.m. (EST). E-mail address: getinfo@ haworthpressinc.com].

minerals. For the 130 years since Constitution, the British Columbia government has endeavored to retain public ownership of its forests, while at the same time trying to ensure rapid exploitation by private firms. This balancing act worked well at first, while there were no concerns about long term sustainability. However as consideration for the future has increased, and the forest resource has become depleted, the balance has become more difficult. In part this is because high interest rates and low growth rates make regeneration in B.C. a poor investment, and because many of the values we derive from the forest cannot be recovered by a private owner. This paper evaluates the effects of recent changes to forestry regulation and operation in B.C., to gauge whether they constitute a substantive move towards a more ecologically and economically sustainable system.

SUSTAINABLE FORESTRY
IN A LOW POPULATION DENSITY SOCIETY

Sustainability issues for North American forests are different to those of South America. Smaller populations and slower growth rates result in less pressure on land. The contested and uncontrollable frontiers between forest and agriculture that exist in South and Central America do not exist. Small areas of Canadian forest are lost each year to agriculture or housing, but most forest land is too far from areas needed for accommodation, or is on terrain unsuited to other uses.

Sustainability should involve maintaining the overall forest ecosystem for posterity. Both the areas of forest and the combination of life forms should be preserved so that if human activity ceased, then the forest would naturally revert to its pre-intruded state. This does not preclude systematic extraction of timber, but does require that no plant or animal species be destroyed either directly or by the destruction of its habitat.

FORESTRY IN BRITISH COLUMBIA

Forest Ecosystems of British Columbia

British Columbia possesses 49 million hectares of non-reserved productive forest, about 21% of such land in Canada. As of 1995, thirty-eight percent of this forest is classified as either being under regeneration or containing immature timber, with sixty-two percent being mature or over-mature. These mature forests are generally now in less accessible areas. The province contains a wide variety of different physiological and biological environments, from the low, warm, wet southern coastal regions to the arid interior grassland and the alpine tundra of the Northeast. The predominant vegetative cover is evergreen coniferous forest. On the coast, western hemlock, western red cedar and Douglas-fir predominate, with Sitka spruce further north. The dry southern interior contains Douglas-fir, Ponderosa pine and larch. Further north, spruce, lodgepole pine and fir are most common. In the southern coastal region and on Vancouver Island there are areas of deciduous forest.

The wide variety of biogeoclimates results in high species diversity, given its northerly location. The 1994 Forest, Range and Recreation Resource Analysis estimates that there are 2,500 native vascular plants and 700 mosses in B.C. Productivity on much Canadian forest land is in an international perspective very low. The average annual forest growth rate for B.C. is 2.3 cubic meters per hectare, while in some South American countries such as Brazil, Chile and Argentina, forests produce up to 70 m^3 per hectare.[1] This low productivity occurs in part because of Canada's climate & soils, and in part because many areas of forest are overmature.

Economic Aspects

British Columbia's economy has traditionally been heavily resource-based, with forestry being the largest sector. The forest industry divides readily into the two sectors of logging and forest products industries. The principal forest products industries are sawmilling, planing mills and shingle mills, veneer and plywood industries and pulp and paper industries, and total value added from these is approximately $6 billion per year. Further direct downstream wood products manufacturing probably add another $1.5 billion. The forest industry forms about 9% of B.C.'s GDP, before considering multiplier effects. About 50% of all exports from B.C. are of timber-related products. The forest industry in total employs about 90,000 people directly, though this is falling steadily as the industry mechanizes and implements new technologies.

The Annual Allowable Cut (AAC[2]) for 1995 on all Provincial and private forest lands (excluding federal lands) in B.C. was seventy-one million cubic meters of both hardwood and softwood, while the actual harvest of roundwood was seventy-seven million cubic meters. The B.C. Forest Act requires that the AAC be met over a five-year period, rather than every year.

FOREST POLICY IN BRITISH COLUMBIA[3]

History

The Land Ordinance of 1865 established the principle of provincial ownership of forest land, and this was extended to B.C. when it joined Confederation in 1871. Although questioned several times by a series of Royal Commissions, the approach has never been changed. Public ownership though has never involved public exploitation. Private firms, who were in a better position to maximize the net benefits from the timber, have always carried out the harvesting and marketing. This split has resulted in persistent incentive problems, the main issue being the absence of motive to regenerate the forest.

Initially the perspective for both government and industry was that there was no scarcity of timber because the forest was so large. Since the harvester had only a short-term lease on the land, that firm would not receive the return if they did make an investment in replanting. Given the assumed surplus of timber and positive interest rates, there was no return to be made. The outcome of this system was that little of the cut area was replanted. There was also no reason for the harvester to minimize damage to the soil or other components of the ecosystem. Irretrievable damage occurred in many areas through soil compaction, the destruction of riparian areas, and erosion on upper slopes.

Before 1887, any conveyance of land automatically included any trees growing on it. From 1896, the definition of 'timber land' came into use, and the sale of such land was forbidden. Some land still passed into private hands with its timber, such as land granted or sold to railways. The first use in B.C. of a timber lease–the right to harvest timber without alienating the land, was in 1865. Several forms of tenure evolved, including Timber Leases and Pulp Leases, which gave the purchaser the right to clear the timber during the life of the lease, with no requirement to regenerate. The Timber Act of 1884 established Timber Licenses, which changed many times over the years, but were essentially short-term renewable harvest licenses. After 1908 these licenses could be transferred to others. These Licenses carried an annual fee, and the harvester also paid royalties on severance. Separate pulpwood licenses were available after 1919.

During the first decade of the twentieth century a very large area of B.C. forest was taken under these various licenses, resulting in a concern that all commercial timber might become committed. The Fulton Commission of 1910 recommended that no further licenses be given, since three-quarters of the province's timber had already been released–more than the forest industry could handle. Over time many of these licenses ended, and subsequent sales were under the 1912 Forest Act that essentially resulted in temporary harvest rights being auctioned off. A Forest Branch was established to conserve the forest, through such measures as fire prevention and reforestation. Logging practices though were highly destructive, leaving large areas, particularly of prime coastal forest, denuded.

In 1945, the Sloan Commission, set up because of concerns about the unbalanced pattern of harvesting and the possibility of an insufficient flow of timber, established in B.C. the concept of sustained yield forest policy. Sloan instituted two types of sustained yield management forms. Tree Farm Licenses (TFLs) enabled the government to impose integrated public management over both existing and new areas of both privately owned and tenured public lands. Licensees have to obtain permits before cutting, and have also to submit management plans every five years. Public Sustained Yield Units were established on land that had not been alienated, and these were to be managed in a similar manner. From 1950 the Forest Service provided seedlings for regeneration on NSR Lands, at no cost. A third Royal Commission reinforced these patterns in 1956.

A number of additional license schemes have been introduced following the 1956 Commission, in part to ensure better utilization of lower grades of wood for pulp and other new products. Silviculture and planting programs were expanded substantially, and by 1968 the Forest Service replanting program had expanded so much that the work had to be contracted out, marking the start of a separate silviculture industry.

In 1976 a further Royal Commission, the "Pearse Commission," reaffirmed the preferability of public ownership of forest land, but observed that this had to be constrained by entrenched private property rights. Pearse therefore did not alter the fundamental structure of property rights, but sought to clarify the objectives of forest policy. His focus was still primarily on the extraction of the resource, at a sustainable rate, rather than on the sustainability of the forest ecosystem itself.

Utilization policies and plans were reorganized, and open competitive bidding was instituted for new "Forest Licenses" (FLs). The period for new TFLs was also standardized to fifteen years, with five year renewals. All former TFLs were converted from indeterminate or perpetual terms, to twenty-one year periods. Various other long-standing but non-perpetual tenures were also changed to fixed terms, with the intention that no forest land should be outside government control for more than twenty-one years.

Because of increased pressure on forest land, it was no longer felt appropriate to allocate any land to exclusive use, even for wilderness. A policy of multiple-use management was formally adopted. Timber harvesting would therefore be carried out as part of this multiple-use, and would no longer dominate the decision-making process. The 1979 *Forest and Range Resource Analysis* predicted a 'falldown' in timber supply of varying magnitude and timing by area, so AACs were cut accordingly. This falldown has now occurred in several areas of the Province.

Public ownership combined with private exploitation of B.C.'s forests is therefore a well-entrenched institution. While there are good reasons for this structure, the results have generally not been sustainable, since the harvester has not had the direct financial incentive to ensure full regeneration of the resource. As it stands, this framework provides inappropriate incentives to the harvester, that result in a systematic failure to regenerate.

Current Property Rights

Throughout the twentieth century the forest resource has been mined rather than managed. Before 1979, when the Ministry of Forests Act was introduced, the explicit intention of the Province's management was to "develop and maintain the province's timber based economy."[4] As time went by more and more land fell from active forest land to "Not Sufficiently Restocked"–vast tracts that had been clearcut and left for possible future regeneration.

The basic pattern of utilization has been for a forest company to obtain a short-term license, during which (subject to limitations) it could remove the trees, and then leave the site. In the past, the company had no responsibility for replanting. Many of the early licenses required the forest company to remove all trees on a site, even those that were not economically worthwhile, which further reducing the potential for natural regeneration. The large scale of most cutblocks also precluded the possibility of tree seeds spreading naturally. Even if natural regeneration was to occur, that would not in most cases result in the immediate restoration of a forest ecosystem similar to that cleared. Complete natural growth to maturity would generally involve several successions of development, with little or no usable timber; recreation or other forest yield during the centuries of evolution. Another frequently occurring problem was the removal of trees from steep slopes high in the watershed, destabilizing upper slopes and leading to rock slides, excessive run-off of precipitation and consequential further erosion.

TIMBER HARVESTING

The general trend in harvest volumes has been upward, particularly in the interior of the Province, with volume peaking in about 1987. Most of the harvest is still taken by the traditional clearcutting approach–more than 90% by the end of the 1980s. Figure 1 shows the annual areas of timber harvested since 1970. The high figures in the 1980s were in part a one-time salvage of trees killed by bark beetles.

Of these areas being cut, only a small proportion was regenerated. Figure 2 shows that during the 1970s less than 40% of the cut area was being replanted.

A direct outcome of this was a steady increase in the area of land that was essentially being left barren to regenerate itself in whatever way occurred 'naturally,' given the absence of both seeds and wind breaks.

Not-Satisfactorily-Restocked Forest Land

The term "Not-Satisfactorily-Restocked Forest Land" (NSR) refers to "forest lands that are not growing to their full timber production potential due to insufficient stocking of acceptable tree species."[5] The failure to regenerate is now being tackled, under three categories. Land from which the trees were removed before 1982 has by law to be reforested by the Ministry of Forests (MoF). The MoF also is required

FIGURE 1. Area of Timber Harvest in B.C. (hectares)

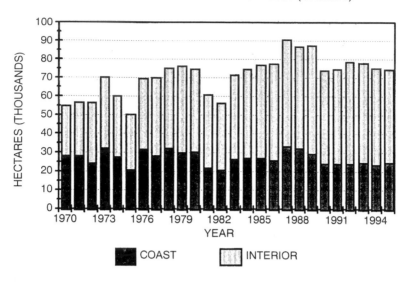

Source: Canadian Council of Forest Ministers, 1996

to regenerate areas cleared between 1982 and 1987, under a different program. After 1987 regeneration became the responsibility of the harvester.

In 1980 there were 1,157,000 hectares of NSR land and also 768,000 hectares of Non-Commercial Brush (NC Brush). At its worst in 1990, almost 8.5% of all forest land in B.C. was in this category, with a further 33% having only immature timber cover. Since 1987 the MoF has cleared a large proportion of what had been a large backlog.

This area has now begun to fall, and the most recent figure, for the fiscal year 1995/96, is 2,964,000 hectares. Some lands have been moved out of the category of NSR, out of the forest reserve, since they are no longer classified as capable of being reforested. Figure 3 shows the pattern of change in NSR land from 1980 to 1996.

POLICY CHANGE BETWEEN 1987 AND 1997

Provincial Ownership of the Forest

Property rights in Canada have, as in most other capitalist democracies, become dichotomized between private and public. Private own-

FIGURE 2. Percentage of Harvest Area Replanted

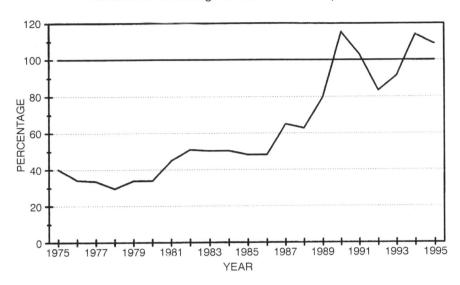

Source: Canadian Council of Forest Ministers, 1996

FIGURE 3. Not Satisfactorily Restocked Forest Land in B.C.

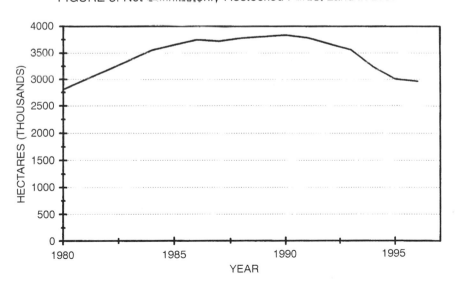

Sources: British Columbia. 1980-1996. *Annual Report of the Ministry of Forests*
Note: These figures include Non-Commercial Brush

ership is not well suited to long-term social and ecological needs of resources such as forests, which generate much value over and above the basic marketable timber, such as recreation, wildlife habitat, carbon sequestration, ecosystem preservation, etc. There is also substantial risk concerning future values. The time for a forest to mature is so great that the financial incentives for a private owner, receiving only the direct value of the timber, are too small to ensure socially appropriate silviculture.

From a social perspective, the probability is that over these very long time horizons some of the non-timber values are likely to appreciate substantially. Some values such as carbon sequestration may become critical to society, yet remain unrewarded for the private owner. Controlling private owners through legislation tends to be unsuccessful due to incentive incompatibility. For these reasons private ownership is inappropriate. Yet pure public ownership has not worked well either. Governments have repeatedly failed to ensure adequate regeneration, and through political pressures have failed to charge realistic severance taxes.

The outcome of the historical system of property rights has been extensive failure to regenerate, because the right to harvest has been logically and legally detached from the requirement to replant. That separation needs to be eliminated, to change the perspective from the natural state of the forest being land, to the view that its natural state is forest. It should never be the case that trees should be removed and land left bare. The non-timber values of a forest are too important. If it is not appropriate to regenerate then it is not logically appropriate to harvest in the first place.

RECENT CHANGES

Forest Management Policy

The first major step in the direction of reorganizing forest management in B.C. came in 1987, when an amendment to the Forest Act made major licensees responsible for the cost of reforestation and basic forestry. This reflected a shift in basic government policy. Major licensees are also required to prepare a "Pre-Harvest Silviculture Prescription," that describes a particular silviculture system and commits them to its use.

In 1989 a Forest Resources Commission was established to "review and make recommendations on the TFL as a form of tenure, clearcut harvesting, public involvement and other important forest management issues."[6] The outcome of this process was a Forest Renewal Plan, implemented in 1994 in response to the impending shortfall in harvests. The plan seeks to boost forest sector employment by enhancing the timber and wood-fibre-using manufacturing sector. The government's intention is to use harvesting rights as a bargaining chip to induce firms to increase employment. A "Timber Supply Review" (TSR) was instigated in 1992, to monitor the state of B.C.'s forests. This involves not just a measurement of the physical characteristics, but their comparison with harvesting and other usage patterns. The principal outcome of the TSR is the AAC, which determines overall harvest volumes.

The *Forest Practices Code of British Columbia Act* (the Code) came into effect in June of 1995, with a phase-in period until June 1997, and exerts jurisdiction over all non-privately owner forest in B.C. Before the Act, there were 20 statutes, 700 sets of separate regulations and many more guidelines, that collectively produced an excessively complicated management structure–complicated to the extent of being in many cases unenforceable. The Code emerged as an outcome of prolonged debate between government, industry and various other groups. There is an element therefore of self-regulation. While the outcome was not fully satisfactory to any of the parties, all had to give up some of their initial demands. The overall effect on industry of the sustainable practices was to increase costs. VanKooten estimates these rose by $3 per cubic meter, whereas Haley puts it higher at $8 per cubic meter.[7] The Code consists of the Forest Practices Act–a legislative umbrella; the regulations; which apply throughout the province; the standards that are amplifications drawn up by the chief forester; and a series of guidebooks. The provisions of the guidebooks are not in themselves mandatory, but the guidebooks provide the basis under which any forest plan is approved.

The natural state of the land is now taken as forest, rather than bare land. Harvest rights are contingent on returning the cut-over area as a twenty-year old forest, with a bond to ensure performance. While retaining ultimate ownership, this new approach shifts the responsibility for forest regeneration onto the harvester. Regeneration of the backlog of NSR and NC Brush land remains to be achieved, but the new property rights structure will end further accumulation of bare land.

Even with the new Act in place, there are several other initiatives in place that affect overall forest sector activity. In 1992 the B.C. government established a Commission on Resources and Environment, that is responsible for developing regional land use planning, and can therefore alter the overall pattern of forestry. There are similar processes at the local level, where the Forest Practices Code is more closely integrated with land-use planning.

The "Canada-British Columbia Subsidiary Agreement" for site rehabilitation and regeneration of NSR lands, which had provided only inconsequential support for regeneration, has been removed. It has been replaced by the Forest Resources Development Agreement of 1985, which allocated $300 million to reforestation of the backlog. A further $200 million was committed in 1991, for direct reforestation and for research purposes. Industry became responsible for much of the forest management that the government had previously undertaken, such as road construction and maintenance.

Impacts

I use here two indicators to trace the effects of the revisions to forestry practices. These are the reductions to the area of NSR land, and changes to the sizes of cutblocks. The rate of regeneration has increased dramatically since 1987, when it became mandatory for all licensees to regenerate the forest in any area harvested. The provincial government also committed to replanting much of the NSR land left over from earlier failures. Other components of silvicultural activity have also increased. Figure 4 shows the net area of all stand tending–which includes all types of silvicultural activity except basic regeneration.

The regeneration rate has increased from around 40% of the harvest area in the 1970s to over 100% in the early 1990s (see Figure 2). This is reflected in the stabilization and then fall in the area of NSR land (see Figure 3). That is due primarily to direct government action in dealing with the backlog of NSR land, rather than regeneration by harvesters. Figure 3 shows not only the land formally classified as NSR, but it also includes land classified as Non-Commercial Brush (NC Brush). Such land has been cut over and not replanted, but currently contains inappropriate growth–brush or trees that will neither generate a commercial crop, nor provide an ecosystem similar to that

FIGURE 4. Area of Stand Tending

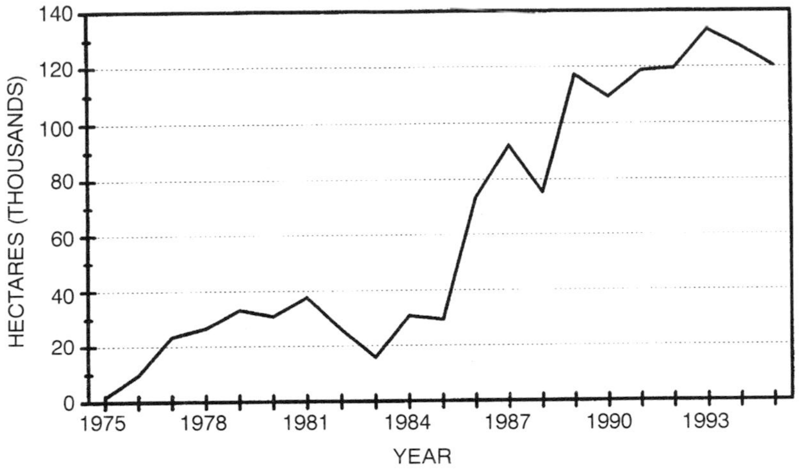

Sources: British Columbia *Annual Report of the Ministry of Forests, 1975-1996.*
Note: These figures include the net area of all types of silviculture excluding basic regeneration (i.e., areas that received more than one operation are included only once).

previously on the site. The high cost of regenerating these sites has meant that the MoF has not been able to tackle much of this area. The total NSR area is unlikely therefore to be eliminated completely.

Figure 5 shows the area of NSR forest land, excluding NC Brush. This has fallen substantially since the inception of the MoF regeneration program in the late 1980s. The MoF is clearing the areas for which it is responsible, but there is a continuing area of land the regeneration of which is the responsibility of the licensee.

The second major change is the manner of cutting. This is being induced primarily by the changes induced by the Forest Practices Code. The National Forestry Database project was initiated in 1990, and the first report on average cut-block size appears in the 1993 report. Table 1 shows the changes in cut-block size since then. The impact of regulation to reduce the size of clearcuts is evident. As harvesting firms are required to undertake full regeneration, the extent of land degradation during cutting is falling. Only the rigid linking of cutting with reforestation can ensure that the incentive to destroy the forest in this way, since the firm doing the felling faces the full cost of restoration.

FIGURE 5. Not Satisfactorily Restocked Areas

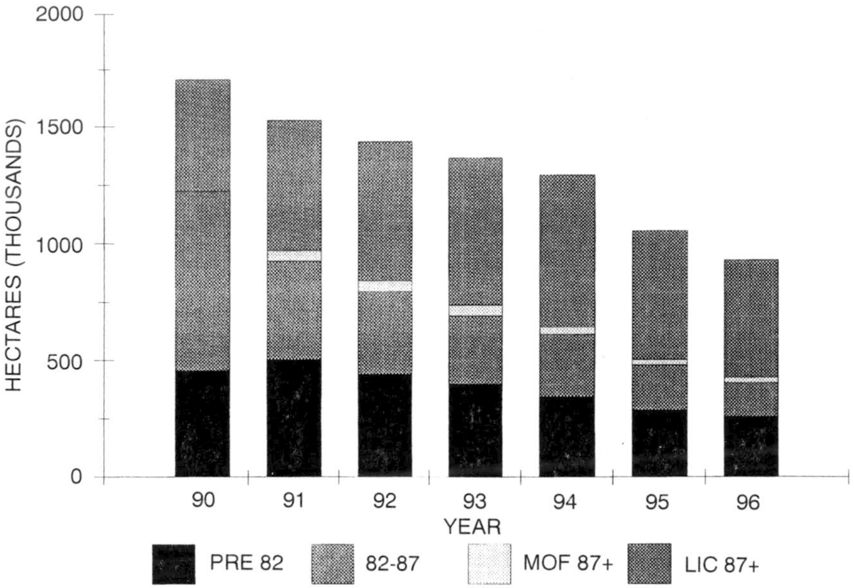

Sources: British Columbia *Annual Report of the Ministry of Forests, 1975-1996.*
Note: No breakdown is available before 1990.
      These areas exclude Non-Commercial Brush.
      Pre 82–land cleared before 1982–The MoF has responsibility for regeneration on these areas.
      82-87–land cleared between 1982 and 1987 again the responsibility of the MoF.
      MoF 87+–land cleared after 1987, but with regeneration the responsibility of the MoF.
      LIC 87+–land cleared after 1987, but with regeneration the responsibility of the licensee.

TABLE 1. Average Cut-Block Size (hectares)

	1993	1995	1996
Cariboo	59.3	43.4	36.6
Kamloops	22.3	21.3	17.6
Nelson	21.3	19.7	16.0
Prince George	60.9	44.4	34.7
Prince Rupert	54.5	31.9	26.5
Vancouver	30.9	31.1	19.1
All Regions	40.2	31.1	25.6

Source: Canadian Council of Forest Ministers. 1997. Compendium of Forestry Statistics 1996.
Note: No data are available before 1993, or for 1994.

CONCLUSION

The various policy initiatives instigated since 1987 are changing forestry practices in B.C. Harvesters are being forced to alter the rates and patterns of both cutting and regeneration. Provincial government ownership is retained, so that over the very long run forest usage remains under political control. The needs of society can therefore be reflected in forest activities. An important outstanding issue in this context is that of the monoculture forest. Regeneration for commercial purposes is generally of just the one tree species that will generate the greatest volume of merchantable timber quickly. This objective may well conflict with the ecologically desirable system for the area. Current regulations do not facilitate the MoF enforcing the more expensive regeneration of a more natural intermixture of species. The timber supply for forest products companies is secure enough for commitments to investment in saw, pulp and paper mills.

These benefits though are achieved at the cost of extensive regulation, which requires all the costs of monitoring and enforcement. Any such structure has inherent instability, with one party experiencing a constant incentive to push at the boundaries of their rights. Clear simple property rights would be preferable from that point of view, but are probably not feasible, given the disparate needs of the various parties. The new Code at least offers the opportunity to enhance sustainability, if the parties adjust their objectives in compliance of the Code.

NOTES

1. Measured as the Mean Annual Increment. Runyon, 1991.

2. The Allowable Annual Cut is now calculated using iterative computer simulations of multiple rotations to explore the effects of changes in flows.

3. This section draws on the 1989 Forest Resources Commission Background Paper on the History of Forest Policy Tenure in B.C., and the 1994 Forest, Range and Recreation Resource Analysis.

4. British Columbia: MoF Annual Report 1994/5, 2.

5. British Columbia. Ministry of Forests. 1994, 2.

6. British Columbia. 1994. *Forest, Range and Recreation Resource Analysis*, p. 284.

7. Cited in Sedjo. (1997), p. 42.

REFERENCES

British Columbia. 1973. A Review of Potential Benefits and Costs Associated with Proposed Rules for Forest Practices in British Columbia. Ministry of Environment, Lands and Parks, Victoria, B.C.

British Columbia. 1971 to 1996. Annual Report of the Ministry of Forests. Ministry of Forests, Victoria, B.C.

British Columbia. 1979, 1984 and 1994. Forest, Range and Recreation Resource Analysis. Ministry of Forests, Victoria, B.C.

British Columbia. 1989. A History of Forest Tenure Policy in British Columbia 1858-1978. Forest Resources Commission, Background Papers, Volume 3, Queen's Printer, Victoria, B.C.

British Columbia. 1994. A Review of Potential Benefits and Costs Associated with Proposed Rules for Forest Practices in British Columbia. Ministry of Environment, Lands and Parks, Victoria, B.C.

Canada. 1990. Regenerating British Columbia's Forests. University of British Columbia Press, Vancouver.

Canada. 1997. The State of Canada's Forests. 1996-1997. Ottawa: Canadian Forestry Service.

Canadian Council of Forest Ministers. 1996. Forest Regeneration in Canada, 1975-1992. Ottawa, Canadian Council of Forest Ministers.

Canadian Council of Forest Ministers. 1997. Compendium of Forestry Statistics 1996. Canadian Council of Forest Ministers, Ottawa.

Cernea, M. 1981. Land Tenure Systems and Social Implications of Forestry Development Programs. Staff Working Paper No. 452, World Bank, Washington.

Drushka, K. W. Nixon and R. Travers. 1993. Touch Wood: B.C. Forests at the Crossroads. Harbour Publishing, Madeira Park, B.C.

Duerr, W.A. 1993. Introduction to Forest Resource Economics. McGraw-Hill, New York.

Englin, J. 1990. Backcountry Hiking and Optimal Timber Rotation. Journal of Environmental Management 31:97-105.

Englin, J. and M. S. Klan. 1990. Optimal Taxation: Timber and Externalities. Journal of Environmental Economics and Management 18: 263-275.

Haley, D. and M. K. Luckert. 1990. Forest Tenures in Canada: A Framework for Policy Analysis. Forestry Canada, Economics Branch, Ottawa.

Hartmann, R. 1976. The harvest decision when a standing forest has value. Economic Inquiry 14: 52-68.

Heaps, T. 1980. Essays on the Qualitative Theory of Forest Economics. Unpublished PhD thesis, University of British Columbia.

Hyde, W. F. 1980. Timber Supply, Land Allocation and Economic Efficiency. Johns Hopkins University Press for Resources for the Future, Baltimore.

Kimmins, J. P. 1994. Sustainable development in Canadian forestry in the face of changing paradigms. The Forestry Chronicle 71: 33-40.

Kimmins, J. P. 1990. Timber Harvesting. Background Paper No. 8. Forest Resources Commission of British Columbia.

Kimmins, J. P. 1997. Balancing Act. Environmental Issues in Forestry. (2nd. ed.) University of British Columbia Press, Vancouver.

Lavender, D. P., R. Parish, C. M. Johnson, et al. 1990. Regenerating British Columbia's Forests. University of British Columbia Press, Vancouver.

Lewis, T. and W. W. Carr. 1989. Developing Timber Harvesting Prescriptions to Minimize Site Degradation-Interior Sites. Ministry of Forests, Timber Harvesting Subcommittee, Victoria, B.C.

Marchak, P. 1983. Green Gold: The Forest Industry in British Columbia. University of British Columbia Press, Vancouver.

Marchak, P. 1995. Logging the Globe. McGill-Queen's University Press, Montreal and Kingston.

North, D. 1990. Institutions, Institutional Change and Economic Performance. Cambridge University Press, Cambridge.

Pearse, P. H. 1976. Timber Rights and Forest Policy in British Columbia. Report of the Royal Commission on Forest Resources, Victoria.

Pearse, P. H. 1988. Property Rights and the Development of Natural Resource Policies in Canada. Canadian Public Policy 14: 307-320.

Redclift, M. 1987. Sustainable Development: Exploring the Contradictions. Methuen, London.

Runyon, K. L. 1991. Canada's Timber Supply: Current Status and Outlook. Forestry Canada, Ottawa.

Sedjo, R. A. (ed.) 1985. Investments in Forestry: Resources, Land Use and Public Policy. Westview, Boulder.

Sedjo, R. A. 1997. The Forest Sector: Important Innovations. Discussion Paper 97-42. Resources for the Future, Washington.

Sloan, Chief Justice Gordon. 1945. Report of the Commissioner relating to Forest Resources of British Columbia. The Government Printer, Victoria, B.C.

Smith, J. B. and D. Tirpak (eds.) 1989. The Potential Effects of Global Climate Change on the United States. United States Environmental Protection Agency.

Sterling Wood Group. 1990. Review of Forest Tenures in British Columbia. Forest Resources Commission, Victoria.

van Kooten, G. C. 1993. Land Resource Economics and Sustainable Development: Economic Policies and the Common Good. University of British Columbia Press, Vancouver.

Walter, G. R. 1980. Financial Maturity of Forests and the Sustainable Yield Concept. Economic Inquiry 18: 327-332.

Deforestation and Land Use Change in Mexico

Juan Manuel Torres-Rojo
Ramiro Flores-Xolocotzi

SUMMARY. This paper addresses the problem of estimating the steady state forest cover in Mexico under different scenarios of production and market conditions. The model is based on the model developed by Ehui et al. (1990), which is adapted to include agriculture and stockbreeding sectors. The model is an optimal control model which assumes a central planning agent who wishes to maximize yields coming from agriculture, livestock and forestry subject to land constraints and dynamics of shifts on land used for different purposes. The model assumes that deforestation is entirely due to changes in land use from forestry to agricultural and stockbreeding purposes. In addition, shifts from land devoted to agriculture to livestock and *vice versa* do not have any effect on forest cover. An empirical application is presented, which uses cross sectional data and resources and market information at State level. Results show that current forest stock can be maintained at interest rates ranging from 3.5 to 4%. Higher interest rates induce deforestation mainly from stockbreeding activities. Finally, increments in value for forest activities have conspicuous effects on mitigating deforestation. *[Article copies available for a fee from The Haworth Document Delivery Service: 1-800-342-9678. E-mail address: <getinfo@haworthpressinc.com> Website: <http://www.haworthpressinc.com>]*

KEYWORDS. Deforestation, dynamic models, steady state forest stock

Juan Manuel Torres-Rojo and Ramiro Flores-Xolocotzi are Associate Professor and Research Assistant, respectively, Center for Research and Teaching of Economics (CIDE), Carr. México-Toluca No. 3655 Col. Lomas de Santa Fé, México D.F. 01210 (E-mail: torrresrj@dis1.cide.mx.).

[Haworth co-indexing entry note]: "Deforestation and Land Use Change in Mexico." Torres-Rojo, Juan Manuel, and Ramiro Flores-Xolocotzi. Co-published simultaneously in *Journal of Sustainable Forestry* (Food Products Press, an imprint of The Haworth Press, Inc.) Vol. 12, No. 1/2, 2001, pp. 171-191; and: *Climate Change and Forest Management in the Western Hemisphere* (ed: Mohammed H. I. Dore) Food Products Press, an imprint of The Haworth Press, Inc., 2001, pp. 171-191. Single or multiple copies of this article are available for a fee from The Haworth Document Delivery Service [1-800-342-9678, 9:00 a.m. - 5:00 p.m. (EST). E-mail address: getinfo@haworthpressinc.com].

171

INTRODUCTION

Approximately 26 percent of Mexico's 191 million hectare (ha) land area (49.7 million ha) is covered with closed forests, most of them temperate forests. However, the country suffers from heavy deforestation problems, especially in the tropics. Some estimates indicate that tropical evergreen forests presently constitute only an estimated 10 percent of their original coverage (Rzedowsky, 1978), and that the whole forest cover is only half the area covered by forests 35 years ago. The problem is that serious that even the 1992 forest global assessment ranked the country in first place given the annual rate of deforestation (1.3 percent) and fourth according to the deforested area (FAO, 1993).

In recent years, deforestation has become a crucial issue in the environmental management agenda for the country. This change of attitudes is due to the society's recognition of the role of forest areas in the production of a number of goods and environmental services. In addition, environmental problems resulting from "El Niño," "La Niña" and global warming, as well as the environmentalist movements (mainly acting on Education, Health and Tourism among others) have forced such a change on the society's perspective. This growing interest on the deforestation problem has forced the most recent change of forest legislation in the country, as well as the increment of budget tied to reforestation, forest health, protection, and sustainable forest management activities.

The most recent forest legislation provides the framework to improve forest activities in a more holistic manner. It considers not only timber production, but also all goods and services produced from forests, as well as the important role of forest communities on forest conservation. This means that logging activities are more constrained and more care is given to conservation and integrated sustainable management practices that consider multiple ecological and socioeconomic factors.

In spite of this new framework, the deforestation process is still taking place and Government actions do not yield the expected results. Commercial harvesting, forest fires and pests contribute to deforest some areas, mainly in the southeast of the country. However, the main cause of deforestation, particularly in the tropics and heavy populated areas, is the conversion of forestlands to agricultural and stockbreeding activities. Expansion of the agricultural frontier and cattle ranching is by far the leading factor in the clearing of forests in Mexico (Toledo, 1990).

Based upon this framework, this paper attempts to estimate deforestation and predict future forest stock conditions based on the assumption that most of the deforestation is due changes in land use. There is a vast literature related to deforestation, most of it seeking to identify causes (Allen and Barnes, 1985), economic problems (Barbier et al., 1991) and policy implications of the process (Dotzauer, 1993). However, much of the economic literature on deforestation is based on econometric models, which stress different factors as causes of deforestation. These factors depend upon the scope of the model. Thus, there are macro-models which explain the process of deforestation for different nations (Barbier et al., 1993; Deacon, 1994). Within this type of models the factors most commonly used to explain deforestation have been population growth, economic development, trade and government policies (Binswanger, 1991). On the other extreme, there exist models which explain deforestation for a particular country or region based on specific factors causing deforestation (Brown and Lugo, 1992; Oyama et al., 1993; Brown and Pearce, 1994). These models refer to variables such as change of land use, technological change, road construction and the market of forest products among others.

The model referred in this paper is an application of the model initially developed by Vousden (1973) and extended by Ehui et al (1990). Additional applications of this model can be found in Ehui and Hertel (1989), and Adamson (1997). The model is an optimal control model, which relies on the estimation of an aggregate production function for the agricultural sector to estimate the desirability of maintaining the forest or clearing the land in a given period. In this paper, the original model is adapted to use cross section data and to include the stockbreeding sector. The paper is organized as follows. The next section shows the basic assumption, the model extension and the main steady state results. The third section presents the results derived from the econometric work and the scenario simulations. Finally, the last section shows some concluding remarks.

MODEL

The original model presents the problem of a central planning agent who attempts to maximize the present value of the utility (social welfare) derived from aggregate profits obtained from the forestry and agriculture sectors (Vousden, 1973). Such a model is adapted to include a third sector, the stockbreeding sector within the same framework.

Model Specification

The objective functional of the model is defined as the maximization of the present value of the utility derived from production in the forestry, agriculture and stockbreeding sectors. The model constraints are defined as changes in forest stock over time and the relationship between yields obtained from different sectors. A fundamental assumption of this model is that forestland can have one additional use that could be agriculture or livestock production, but not both additional uses. In other words, there is an initial endowment of forestland, which can have either of the two sets of uses: agriculture and forestry or forestry and stockbreeding, but not agriculture, forestry and stockbreeding. Indeed, this last possibility of potential land use is feasible, however forestland is converted to just one of the two additional uses. In the long run, this possibility establishes shifts in land use from agriculture to stockbreeding and *vice versa*. Such shifts, although possible, are just part of the land accounts and do not affect forest cover.

Formally, the control problem is stated as:

$$\underset{D_A, D_L, X_A, X_L}{Max} \quad U = \int_0^{\infty} [W(\Pi(D_A, D_L, X_A, X_L, F))]e^{-rt} dt \tag{1}$$

subject to

$$
\begin{aligned}
F_A &= -D_A(t) \\
F_L &= -D_L(t) \\
F_A(t), F_L(t), D_A(t), D_L(t), X_A(t), X_L(t) &\geq 0 \\
F_A(0) + F_L(0) &= F(0) = L_A + L_L = given \\
D_A(t) + D_L(t) &= D(t)
\end{aligned} \tag{2}
$$

$$
\begin{aligned}
\Pi(\cdot) = \ & P_F(t) F(t) + [L_A - F_A(t)]\{P_A(t) Z [D_A(t), F_A(0) - F_A(t), X_A(t), \\
& PP(t)] - P_{XA} X_A(t)\} + [L_L - F_L(t)]\{P_L(t) Y [D_L(t), F_L(0) \\
& - F_L(t), X_L(t)] - P_{XL} X_L(t)\}
\end{aligned} \tag{3}
$$

Where U represents the present value of society's welfare, r is the social rate of discount; $W(.)$ is a twice differentiable welfare function which depends upon aggregate profit Π. The profit function (3) is the sum of net returns obtained from forestry, agriculture and stockbreeding; L_A represents the total land area suitable for agricultural use, while L_L is the total land suitable for livestock production. $F(t)$ represents the total land area covered by forests at time t, while $F_A(t)$ is the land area currently covered by forests that can be used for agricultural purposes in contrast to $F_L(t)$, that represents the forested area that might be used for livestock production. It is assumed that the forestland can be used either for agricultural or livestock production, and there is some forestland that can be used exclusively for forestry.

$Z(\cdot)$ is a concave yield function for the agricultural sector, while $Y(\cdot)$ represents the (concave) yield function for the stockbreeding sector. Both production functions depend upon the purchased inputs ($X_A(t)$ for the agricultural sector, and $X_L(t)$ for the stockbreeding sector). $D(t)$ represents the total rate of deforestation, which can be due to the agricultural sector ($D_A(t)$) or to the livestock sector ($D_L(t)$). The difference $[F_i(0) - F_i(t)]$ shows the cumulative amount of deforested land to be incorporated to the i-th sector, while $PP(t)$ represents annual precipitation at time t. Variables $P_A(t)$ and $P_L(t)$ denote prices per unit (e.g., Kg, Tn) returns to agriculture and stockbreeding at time t. $P_F(t)$ denotes per hectare returns to forestry at time t. Following the same nomenclature, $P_{XA}(t)$ denotes per unit agricultural input prices, while $P_{XL}(t)$ indicates per unit prices for inputs used in livestock production.

Model Assumptions

The formulation presented above has the following characteristics and assumptions:

i. The total forestland can be used either for agricultural or stockbreeding production. However, the forestland incorporated into the agricultural sector cannot be incorporated also into the livestock production after it has been cleared. Shifts of land between the agricultural and stockbreeding sectors might exist and the model does not account for such changes. This assumption is somehow restrictive in the short run; however, in the long run land should be allocated to its most profitable use. The experience in Mexico shows that it is very unlikely that land might be returned to the forestry sector in the future once it is cleared (SARH, 1990), unless some land conversion or land rehabilitation governmental program is conducted. For this reason, this assumption does not affect the main purpose of the model, namely, to evaluate deforestation.

ii. Social welfare increases as profit increases and there exist diminishing marginal utilities from additional profits. Formally, this assumption implies:

$$\partial W[\Pi]/\partial \Pi \geq 0$$
$$W_{\Pi \Pi} \leq 0$$

iii. Agricultural yields increase as deforestation increases but at a decreasing rate. This effect is attributable to the nutrient content of the ashes left after burning, especially in tropical forests (Sánchez, 1976; Hernández et al., 1987; Ehui et al., 1990; Levy et al., 1991). Such ashes suffer a fast degradation process, leaving soils with a low nutrient content. In the case of temperate forests, a similar effect can be attributable to bare land erosion and lack of soil conservation practices after clearing or burning (Sánchez and Ortiz, 1991). This assumption implies:

$$\partial Z/\partial D_A (t) > 0$$
$$\partial^2 Z/\partial D^2_A (t) < 0$$
$$\partial Z/\partial [F_A (0) - F_A (t)] < 0$$

Observe that the third inequality assumes that for a given period t, as more marginal lands (lands currently covered by forests) are incorporated into the arable land base, lower agricultural yields are obtained. This assumption is consistent with the fact that current forestlands are located on steep terrain, high altitudes and with poor infrastructure for access.

iv. Forestlands incorporated into stockbreeding production are only used for cattle raising. It is assumed that this activity only affects the stock of goats, sheep and cattle, which are used exclusively for meat production. In addition, it is assumed that increments of deforestation increase meat production but at decreasing rates, because of increases in costs associated with ranching more distant areas as well as costs associated with crowding (Aguirre, 1996). This assumption formally implies:

$$\partial Y/\partial D_L (t) > 0$$
$$\partial^2 Y/\partial D^2_L (t) < 0$$
$$\partial Y/\partial [F_L (0) - F_L (t)] < 0$$

Observe also that the last inequality indicates that the use of forest-lands for grazing leads to lower meat yields, implying that the forest-land is a marginal land for livestock production; this assumption might not hold in some cases.

v. Agricultural and meat yields increase as inputs increase but at diminishing marginal yields. This assumption results from the basic idea of diminishing marginal yields from any resource, an assumption also valid for both biological production systems: agriculture and stockbreeding. The assumption thus implies:

$$\partial Y/\partial X_L\ (t) > 0 \qquad \partial Z/\partial X_A\ (t) > 0$$
$$\partial^2 Y/\partial^2 X_L\ (t) < 0 \qquad \partial^2 Z/\partial^2 X_A\ (t) < 0$$

vi. The following second order derivatives are supposed to equal zero in order to facilitate getting some of the results:

$$\partial^2 Y/\partial^2 F_L\ (t) = 0 \qquad \partial^2 Y/\partial F_L\ (t)\partial X_L\ (t) = 0$$
$$\partial^2 Z/\partial^2 F_A\ (t) = 0 \qquad \partial^2 Z/\partial F_A\ (t)\partial X_A\ (t) = 0$$

Steady State Solution

The current value Hamiltonian associated with model describe by (1)-(3) is given by:

$$\tilde{H} = W(\Pi\ (D_A, D_L, X_A, X_L, F)) - \lambda D_A\ (t) - \mu D_L\ (t) \qquad (4)$$

Where λ and μ denote current value costate variables associated with the equations of motion defined in (2). Assuming the existence of an interior solution, the maximum principle requires the following conditions to hold:

Optimality conditions:

$$\tilde{H}_{XA} = W_\Pi\ [(L_A - F_A\ (t))\ (P_A\ Z_{XA} - P_{XA})] = 0 \qquad (5)$$

$$\tilde{H}_{XL} = W_\Pi\ [(L_L - F_L\ (t))\ (P_L\ Z_{XL} - P_{XL})] = 0 \qquad (6)$$

$$\tilde{H}_{DA} = W_\Pi\ [(L_A - F_A\ (t))\ (P_A\ Z_{DA} - P_{XA})] = \lambda \qquad (7)$$

$$\tilde{H}_{DL} = W_\Pi\ [(L_L - F_L\ (t))\ (P_L\ Y_D - P_{XL})] = \mu \qquad (8)$$

Costate equations:

$$-\tilde{H}_{FA} = \dot{\lambda} - r\lambda = W_\Pi \left[P_F + (L_A - F_A(t))(P_A Z_{FA}) - (P_A Z) + (P_A X_A)\right] \quad (9)$$

$$-\tilde{H}_{FL} = \dot{\mu} - r\mu = W_\Pi \left[P_L + (L_L - F_L(t))\right.$$
$$\left.(P_L Y_{FL}) - (P_L Y) + (P_L X_L)\right] \quad (10)$$

and transversality conditions given by:

$$\lim_{t \to \infty} e^{-rt}\lambda(t)F_A(t) = 0 \quad (11)$$

$$\lim_{t \to \infty} e^{-rt}\mu(t)F_L(t) = 0 \quad (12)$$

Conditions (5)-(10) have the same interpretation as the one given by Ehui and Hertel (1989). Conditions (5) and (6) imply that at the optimum the purchased inputs have to be applied at the level where marginal utilities are zero. Equation (7) indicates that at any point in time, the rate of deforestation coming from the agriculture sector should be chosen so that the marginal utility of deforestation equals the opportunity cost of the forest stock (λ). Condition (8) has an analogous interpretation as equation (7) but for the stockbreeding sector. Conditions (9) and (10) imply that forest stock should be employed to the point where marginal utility of forest capital is equal to the social cost of such a capital. Observe that in these conditions the right hand side of equations (9) and (10) integrate both: the marginal contribution of forestry ($W_\Pi P_F$) and the indirect marginal contribution of the forestland to agricultural and stockbreeding productivity.

Under steady state conditions, the change on forest stock should be zero, which means that:

$$\dot{F}(t) = \dot{F}_A(t) = \dot{F}_L(t) = D_A(t) = D_L(t) = D(t) = 0 \quad (13)$$

By using this assumption and equations (5)-(8) the steady state forest stock can be defined by:

$$\frac{1}{r}W_{FA}(D_A, F_A, X_L, X_A) = W_{DA}(D_A, F_A, X_L, X_A) \quad (14)$$

$$\frac{1}{r}W_{FL}(D_L, F_L, X_L, X_A) = W_{DL}(D_L, F_L, X_L, X_A) \quad (15)$$

$$Z_{XA}(D_A, F_A, X_A) = \frac{P_{XA}}{P_A} \quad (16)$$

$$Y_{XL}(D_L, F_L, X_L) = \frac{P_{XL}}{P_L} \quad (17)$$

Since by assumption (i) there is no land that can be potentially used for agriculture, forestry or stockbreeding, the first order conditions for agriculture (5 and 7) are totally independent from the ones for stockbreeding (6 and 8). Therefore, conditions (14–17) are basically an extension of those developed by Ehui et al. (1990). Condition (14) states that in steady state, the marginal utility of deforestation made on forest lands with likely agricultural use (W_{DA}) must equal the present value of the foregone marginal future benefit of those lands $\left(\frac{1}{r}W_{FA}\right)$.

Condition (15) just indicates the same argument for those forestlands with likely use for livestock production. Ehui et al. (1990) called (W_{DA}) the preference for deforestation (in this case due to change of land use for agriculture purposes) and considered (W_{FA}) the conservation motive. Such a terminology is also valid for this model extension given separate sources of deforestation. Equations (16) and (17) just show the basic equilibrium condition in production theory: the value of the marginal product equals the marginal cost for both aggregate products (see Ehui et al., 1990, for a detailed discussion on these conditions and the associated phase diagrams).

Specification of Yield Functions

The basic idea of the yield functions is to estimate an aggregate yield for both sectors, namely agricultural sector $[Z(t)]$ and the stockbreeding $[Y(t)]$ sector. Because second order derivatives of yield functions for some parameters are critical to the analysis, the second order approximation developed by Ehui and Hertel (1989) was adopted with few changes. The quadratic functional form adopted for the agricultural production function was:

$$Z(t) = a_0 + a_1[X_A(t)] + a_2[D_A(t)] + a_3[F_A(0) - F_A(t)] + a_4[AT(t)]$$
$$+ \frac{1}{2}a_{11}[X_A(t)]^2 + \frac{1}{2}a_{22}[D_A(t)]^2 + a_{12}[D_A(t)X_A(t)] \quad (18)$$

Where the term $AT(t)$ denotes technological change in the agricultural sector. The functional form adopted for the livestock production sector was:

$$Y(t) = \beta_0 + \beta_1[X_L(t)] + \beta_2[D_L(t)] + \beta_3[F_L(0) - F_L(t)] + \beta_4[LT(t)]$$
$$+ \beta_5[PP(t)] + \beta_6[D_Y] + \frac{1}{2}\beta_{11}[X_L(t)]^2 + \frac{1}{2}\beta_{22}[D_L(t)]^2$$
$$+ \beta_{12}[D_L(t)X_L(t)] \quad (19)$$

For this model the term $LT(t)$ denotes technological change in the stockbreeding sector at time t. The term $PP(t)$ represents the precipitation at time t and D_Y is a dummy variable to account for different time periods in the intercept. AT is measured in terms of machinery per unit of arable land, while LT is measured by the liters of milk produced, since this production reflects a proxy of the stabled stock of cattle.

As can be observed, both equations are not complete second order approximations of the functional form, since only important interactions and quadratic terms are considered. The main reason for ignoring some terms is the lack of adequate information. Only one interaction term is included, $X(t)$ and $D(t)$ because according to Sánchez (1976, 1991) current period deforestation is analogous to fertilization, whose effect decreases rapidly after some years. Hence this term enables us to test this statement.

Considering the assumptions about the yield function the following signs are expected for the parameters in both equations: $\alpha_0, \alpha_1, \alpha_2, \alpha_4,$ $\alpha_{25}, \beta_0, \beta_1, \beta_2, \beta_4, \beta_5 \geq 0$; $\alpha_3, \alpha_{11}, \alpha_{22}, \alpha_{12}, \beta_3, \beta_{11}, \beta_{22}, \beta_{12} \leq 0$.

Steady State Forest Cover

By solving equations (14) to (17) the optimal steady state forest stock can be found. The optimal level of inputs X^* can be found by solving equations (16) and (17). For the agricultural sector the resulting levels are:

$$X_A^* = \frac{(\overline{P}_{XA} - \alpha_1)}{\alpha_{11}}$$

where $\overline{P}_{XA}$ represents the ratio P_{XA}/P_A. Similarly, by defining $\overline{P}_{XL}$ as the ratio P_{XL}/P_L the optimal input level for the stockbreeding sector yields:

$$X_L^* = \frac{(\overline{P}_{XL} - \beta_1)}{\beta_{11}}$$

By solving equation (14) and considering (13) optimal forest cover with possible agricultural use yields:

$$F_A^* = F_A(0) + \frac{\Delta}{\Omega} + \frac{(\Omega - \alpha_3)A}{\Omega}$$

Where:

$$\Delta = \alpha_0 + \alpha_1 X_A{}^* + \tfrac{1}{2}\alpha_{11}X_A{}^* + \alpha_4 AT - \overline{P}_{XA}X_A{}^* - \overline{P}_{FA}, \text{ and}$$
$$\Omega = r(\alpha_2 + \alpha_{12}X_A{}^*) + 2\alpha_3$$

Here $\overline{P}_{FA}$ represents the per hectare returns from forestry activities relative to the price of agricultural outputs P_F/P_{XA} and $A = L_A - F_A(0)$. Following the same steps and by using equation (15), the optimal forest cover of forestlands that can also be used for livestock production is:

$$F_L{}^* = F_L(0) + \frac{\Gamma}{\Phi} + \frac{(\Phi - \beta_3)B}{\Phi}$$

Where:

$$\Gamma = \beta_0 + \beta_1 X_A{}^* + \tfrac{1}{2}\beta_{11}X_A{}^* + \beta_4 LT + \overline{P}_{XL}X_L{}^* - \overline{P}_{FL}, \text{ and}$$
$$\Phi = r(\beta_2 + \beta_{12}X_L{}^*) + 2\beta_3$$

For these equations $\overline{P}_{FL}$ represents the per hectare returns from forestry activities relative to the price of livestock outputs P_{FL}/P_{XL} and $B = L_L - F_L(0)$. Comparative static results of the model are detailed in Ehui and Hertel (1989) and Ehui et al. (1990).

EMPIRICAL APPLICATION

The following application was carried out by using data from Mexico. Given that only three forest inventories have been performed in the country, the main limitation was the availability of forest data. Hence the information was grouped for each one of the states and runs to estimate agricultural and stockbreeding production functions were performed at state level.

Database

Forest cover estimates for Mexico depend upon the source of information and the method used to estimate and classify forest vegetation.

Cairns et al. (1995) have found that different sources of information have different definitions of forest types and degrees of forest degradation. Most of them use different methodologies to sample and estimate forest cover (Castillo et al., 1989; FAO, 1990; Masera et al., 1992; SARH, 1986). There is only one source of information that provides the most complete data under a consistent format. This source is the state by state forest inventories conducted in the country by the National Forest Service. The first state level forest inventory was initiated at the beginning of the sixties and finished in the middle of the 80s (SARH, 1986). A second forest inventory was performed in 1990 (SARH, 1992) and the third one was performed in 1994 (SARH, 1994). These three nation wide forest inventories integrate the only information available of forest cover for the whole country.

Given the lack of reliable time series information at national level, a different approach to estimate the production functions was followed. Instead, cross section information at state level was used to fit the models. Thus, the sample consisted of 31 states (the federal capital was not included) and two measures of deforestation were used: one obtained from the first and second forest inventories and the second one by using the second and third forest inventories. Price information from the forest sector was obtained from the quarterly (and sometimes monthly) economic reports from the National Forest Service. Mexican forest industry outlook reports were also used to gather some additional information.

Aggregate agricultural yields were estimated according to a quantity index for the major food crops produced in marginal lands in Mexico. These crops included: rice, corn, wheat, barley, beans, potato and sorghum. The quantity index was estimated as the geometric mean of the quantity indexes estimated as:

$$I_{90} = \frac{\sum_{j} p_{jtk}\, q_{jtk}}{\max_{k}\left(\sum_{j} p_{j90k}\, q_{j90k}\right)} \qquad I_{94} = \frac{\sum_{j} p_{jtk}\, q_{jtk}}{\max_{k}\left(\sum_{j} p_{j94k}\, q_{j94k}\right)}$$

Where I_{90} and I_{94} are the quantity indexes based on prices indexed to 1990 and 1994, respectively. The values p_{jtk} and q_{jtk} are, respectively the price and quantity of the j-th commodity, at time t (indexed according to the base year defined by the index, e.g., 1990 or 1994) produced in the state k. The production quantity index (I) was then estimated as:

$$I = \sqrt{I_{90} \cdot I_{94}}$$

Agricultural production, market information on products and factors at state level were obtained from the FAO/IMTA (1995) agriculture database, as well as SARH (1980) and SAGAR (1995) database. Additional information was obtained from the INEGI's agricultural census and the Ministry of Agriculture bulletins (SARH, 1993). Estimates about potential use of forest lands was obtained from different sources such as SARH (1993) and SAHOP (1991).

Land devoted to agriculture was used as a proxy for the agricultural inputs. The land was weighted by the corn yield obtained for each one of the states under two classes: irrigated and non-irrigated land. Then a new variable expressing the total agricultural land was obtained as:

$$X_A = \frac{y_i * L_i + y_n * L_n}{y_n}$$

Where y represents the corn yield either in irrigated land (subscript i) or in non-irrigated land (subscript n) and L is the total amount of land devoted to agricultural production either in irrigated land (subscript i) or in non-irrigated land (subscript n).

Aggregate meat yields were also measured (in heads of cattle) through indexes from cow, sheep and goat heads produced in Mexico (at state by state basis). Meat production, number of heads, productivity indexes for forage and price information for the stockbreeding sector were obtained from SARH (1994), SAGAR (1996) and SAGAR (1998).

An estimate of amount of forage obtained from land devoted to grazing was calculated by using the information on forage production and grazing productivity published by COTECOCA (1994) and SAGAR (1998). This total amount of forage at state level was used as a proxy for the input level to estimate the production function (19). Annual precipitation was considered another important input for the production function. It was obtained from the database published by Quintas (1996).

Returns were defined as follows. Annual per hectare returns in agriculture was defined as the total annual real value of the food crops listed above divided by the area used to produce them. Annual per hectare meat production returns were defined as the total annual real value of the meat produced by the standing heads (cows, sheep, and goats) divided by the number of hectares of grazing. Finally, annual per hectare returns to forestry were defined as the weighted (by volume produced) average annual price of all timber species times their average annual production per unit of land (in cubic meters per hectare). This latter result is obtained by multiplying the density of tradable trees (in cubic meters per hectare) by their annual growth rate.

Estimated Models

Estimates for the yield functions $Z(t)$ and $Y(t)$ were obtained by using Weighted Least Squares estimation. Production indexes were normalized by using a Box-Cox power transformation. Resulting estimates are:

$$Z(t) = 2.2961 + 0.001896\ X_A\ (t) + 0.0263\ D_A\ (t) - 0.0032[F_A\ (0) - F_A\ (t)]$$
$$\quad (7.288) \qquad (2.584) \qquad\qquad (1.027) \qquad\qquad\qquad (-2.133)$$
$$+\ 1.30075\ AT - 0.363\ X_A{}^2(t) - 0.000237\ D_A{}^2(t) - 0.01737\ [X_A\ (t)*D_A\ (t)]$$
$$\quad (3.361) \qquad (-0.830) \qquad\quad (-2.330) \qquad\qquad (-0.897)$$
$$R^2 = 0.662 \qquad F = 13.59 \qquad DW = 1.967 \qquad n = 62$$

and

$$Y(t) = 0.800263 + 0.001462\ X_L\ (t) - 0.007364\ D_L\ (t) + 0.000354[F_L\ (0) - F_L\ (t)]$$
$$\quad (2.685) \qquad (5.032) \qquad\qquad (-2.325) \qquad\qquad\qquad (1.789)$$
$$+\ 1.7962\ LT - 0.2581\ X_L{}^2(t) + 6.8341\ D_L{}^2(t) + 0.97648\ [X_L\ (t)*D_L\ (t)]$$
$$\quad (4.679) \qquad (-3.638) \qquad\quad (1.701) \qquad\qquad (1.953)$$
$$+\ 0.000461\ PP - 0.5732\ Y_L$$
$$\quad (2.259) \qquad\quad (-2.821)$$
$$R^2 = 0.7682 \qquad F = 14.732 \qquad DW = 1.914 \qquad n = 62$$

In both equations, numbers in parenthesis show t-values for the parameter estimate right above. All variables keep the same meaning as defined in the last section.

As can be observed the agricultural yield function yielded the expected signs for all the estimates. However, the estimates for deforestation (D_A) and the square of inputs ($X_A{}^2$) were not statistically significant. The statistical significance for $D_A{}^2$ confirms that deforestation increases agricultural production at a decreasing rate. On the other hand, the poor significance for the $X_A{}^2$ might suggest that the agricultural production function is not concave. However, this result is likely due to the interval of values used for model fitting and the way the inputs are estimated.

The production function for the stockbreeding sector shows several estimates with signs different from expected ones. For instance, the estimates for D_L and $D_L{}^2$ show that the production function in the interval considered by the data set is convex with respect to deforestation. In other words, deforestation increases at increasing rates with stockbreeding production. In concordance with this behavior, the estimate for $[F_L(0) - F_L(t)]$ is positive, meaning that the forest lands incorporated for stockbreeding production in fact are not marginal lands, but they increase production in this sector.

Simulation Results

Once yield functions were estimated, the so called "socially optimal" steady state forest stock (F^{**}) was computed by summing for all states the estimated optimal state level forest stocks from the forest land with likely agricultural use (F_A), and the forest land with likely stockbreeding use (F_L). These computations were accomplished by defining some values for the technological levels (same as the 1994s conditions), precipitation (average precipitation in the last 25 years) and by assuming that Y_L equals zero.

Table 1 shows the results obtained when different values for the prices of agricultural products (P_A) and real interest rate (r) are assumed. The column named original values shows the expected forest

TABLE 1. Forest stock given different price rations in agricultural production

r	Original values	$2P_A$	$4P_A$
0.03	15080.132	14153.9998	11812.3216
0.05	12920.4475	10663.8107	3186.9745
0.07	9877.9857	4774.975	− 25772.7836
0.09	5272.2308	− 7286.4608	−
0.11	− 2519.229	− 46063.9709	−

stock given the initial price relationship, while the following columns show the expected forest stock (with potential agricultural use) when the price for agricultural outputs is doubled ($2P_A$) and quadrupled ($4P_A$).

Table 2 shows the results obtained from simulations when different values for the prices in the stockbreeding sector (P_L) and real interest rate (r) are assumed. As in Table 1, the column named original values shows the expected forest stock (with potential stockbreeding use) given the initial price relationship, while the following columns show the expected forest stock when the price for stockbreeding outputs is doubled ($2P_L$) and quadrupled ($4P_L$).

Observe that in Table 1 and Table 2, forest stock shows only the area covered by forest with potential agricultural or stockbreeding production. In order to estimate the total forest stock, values in both tables must be combined.

These results show that interest rate has stronger effect on shifting forestland to stockbreeding production than on shifting forestland to agricultural production. This is likely due to the fact that livestock production yields higher returns than agricultural production per unit of land area, especially in forestlands, which happen to be marginal for agriculture but not for stockbreeding. The effect on prices is quite different, since a price increment in the agricultural sector has a stronger effect on reducing forest cover than the same proportional price increment in the stockbreeding sector. The result might be explained by the fact that cattle ranching is a very extensive activity, and hence an increment in prices which might be interpreted as a short run effect, does not change the stockbreeding production area. On the other hand, agriculture, which is an annual activity and more intensive than cattle ranching, does have a stronger impact as a result of a change in prices.

The policy implication from these results is that an incentive pro-

TABLE 2. Forest stock given different price ratios in livestock outputs

r	Original values	2 PL	4 PL
0.03	53565.921	53564.315	53561.101
0.04	38714.727	38712.834	38709.048
0.05	17426.557	17424.255	17419.65
0.06	− 15641.449	− 15644.388	− 15650.266
0.07	− 74006.858	− 74010.92	− 74019.044

gram in the agricultural sector is more likely to produce more deforestation than and incentive program in the stockbreeding sector. In addition, the stabilization of real prices results in a fast reduction on the desirability to convert forestlands into grazing fields.

The next question is how sensible is the forestland to changes in forest products prices. One might expect that increasing forest products prices increase drastically forest cover. Table 3 shows these sensitivities when prices for forest products are doubled (*2F*) or quadrupled (*4F*). Notice the differential effect from the agricultural and stockbreeding sectors which is consistent with the result that marginal rate of transformation is larger for the stockbreeding and forestry sectors than for agriculture and forestry. Anyhow, increments in forest products prices have conspicuous effects on reducing deforestation.

The final parameter tested in the simulations was technological level. Table 4 and Table 5 show the effect of technology in the change of land use from forestry to agriculture and stockbreeding. Results show almost no effects from increments in technology level (either agriculture or stockbreeding) on changes on land use. This result just confirms that by increasing the productivity of current agricultural and stockbreeding land, it is possible to reduce deforestation induced by market causes.

Finally, all tables show that current forest stock conditions can be

TABLE 3. Forest stock given different price ratios for forest products

r	F_A P_F	F_L P_F	$(F_{L+} F_A)/P_F$	$F_A/2P_F$	$F_L/2P_F$	$(F_{L+} F_A)/2P_F$	$F_A/4P_F$	$F_L/4P_F$
0.03	15080.13	53565.92	68646.05	15080.14	53566.72	68646.87	15080.17	53567.11
0.05	12920.44	17426.55	30347.00	12920.46	17427.70	30348.17	12920.50	17428.23
0.07	9877.98	−74006.85	−64128.87	9878.01	−74004.82	−64126.81	9878.05	−74004.02

TABLE 4. Forest stock given different values in technology level of agriculture

r	Original values F_A	2 AT	4 AT
0.03	15080.132	15077.8221	15073.2024
0.05	12920.4475	12917.7468	12912.3454
0.07	9877.9857	9874.735	9868.2335
0.09	5272.2308	5268.1486	5259.9842
0.11	−2519.229	−2524.7157	−2535.6872

TABLE 5. Forest stock given different values in technology level in stockbreeding

r	Original values F_L	2 LT	4 LT
0.03	53565.921	53554.403	53548.651
0.04	38714.727	38701.147	38694.37
0.05	17426.557	17410.018	17401.773
0.06	− 15641.449	− 15662.598	− 15673.045

conserved at an interest rate ranging from 3.5%-4.0%. Market conditions with interest rates above 5.5% lead to deforestation of current forest stock.

CONCLUSIONS

The results of this paper show the expected behavioral change due to land use; the larger the interest rate, the smaller the conservation motive and the smaller the amount of forest that will be conserved. On the other hand, the greater the prices for agriculture and meat products, the greater the benefits from the change of forestland use which yield smaller forested areas. An important result derived from this analysis is that the stockbreeding sector deforest forestlands faster than the agricultural sector. In addition, forestlands converted to stockbreeding remain highly productive at least for the time interval analyzed. On the contrary, forestland converted to agriculture remains highly marginal. This means that the stockbreeding sector is the one causing most of the reduction of the conservation motive, since it makes it more desirable to convert forestlands.

Unfortunately for forestry, increments in the value of forest products (produced in the forest) do not have a strong effect on mitigating deforestation. That means that additional activities which might increase the per hectare value of forestry such as use of non-timber forest products, hunting, and recreation, among others, have almost no effect on reducing deforestation.

One important extension of the model should be to evaluate the effect of policy reforms on the agriculture sector. These reforms consider the use of subsidies per unit of arable land for some crops as well as trading subsidies. These subsidies obviously will reduce the conservation motive and lower the desirability of maintaining the forest stock. A more complete model should be developed to evaluate the impact of excessive deforestation rates on agriculture and livestock production by integrating the externalities and additional costs (caused by deforestation), directly into the production functions.

REFERENCES

Aguirre, O. 1996. Dinámica de los agostaderos del Estado de México. Estudio Especial de Agostaderos. Programa de Modernización del SEDEMEX. PROBOSQUE. Metepec, Estado de México.

Adamson, M. 1997. Deforestación, producción agrícola y ganadería en Costa Rica. *In*: Calvo, E. Figueroa and J. Vargas Eds. Medio ambiente en Latinoamérica: Desafíos y propuestas. Univ. de Costa Rica, Univ. de Chile. San José, Costa Rica, pp. 177-216.

Allen, J.C. and D.F. Barnes. 1985. The causes of deforestation in developing countries. *Annals of the Association of American Geographers* 75(2):163-184.

Barbier, E.B., J.C. Burgess and A. Markandya. 1991. The economics of tropical deforestation. *Ambio* 20(2):55-58.

Barbier, E.B., J.C. Burgess, J. Bishop, B. Aylward and C. Bann. 1993. The economic linkages between the international trade in tropical timber and the sustainable management of tropical forests. ITTO Final report.

Binswanger, H.P. 1991. Brazilian policies that encourage deforestation in the Amazon. *World Development* 19(7):821 829.

Brown, S. and A.E. Lugo. 1992. Aboveground biomass estimates for tropical moist forests of the Brazilian Amazon. *Interciencia* 17(1):8-18.

Brown, K. and D.W. Pearce (Editors). 1994. The causes of deforestation: The economic and statistical analysis of factors giving rise to the loss of tropical forests. UBC Press. Vancouver B.C., Canada.

Cairns, M.A., R. Dirzo and F. Zadroga. 1995. Forests of Mexico. *Journal of Forestry*, 93(7)21:24.

Castillo, P.E., P. Lehtonen, M. Simula, V. Sosa C. and R. Escobar. 1989. Proyecciones de los principales indicadores forestales de México a largo plazo (1988-2012). Internal report. Cooperation Project Mexico-Finland. Subsecretaría Forestal, SARH, México.

Cavazos D., J.R. 1997. Uso múltiple de los agostaderos en el norte de México. *Ciencia Forestal* 22(81):3-26.

Comision Tecnica Consultiva de Coeficientes de Agostadero (COTECOCA). 1994. Revegetación y reforestación de las áreas ganaderas en las zonas áridas y semiáridas de México. Mexico city. SARH.

Deacon, 1994. Deforestation and the rule of Law in a cross-section of countries. *Land Economics* 70(4):414-430.

Dotzauer, H. 1993. Los factores políticos y socioeconómicos que causan degradación forestal en la República Dominicana. Rural Development Forestry Network. Documento 16d., 20 p.

Ehui, S. and T. Hertel. 1989. Deforestation and agricultural productivity in the Cot d'Ivory. *American Journal of Agricultural Economics* 71(3):703-711.

Ehui, S., T.W. Hertel and P.V. Preckel. 1990. Forest resource depletion, soil dynamics and agricultural development in the tropics. *Journal of Environmental Economics* 18(1):5-29.

Food and Agriculture Organization (FAO). 1993. Forest resources assessment 1990. Tropical countries. Forestry paper No. 112. Rome, FAO.

Food and Agriculture Organization (FAO)/Instituo Mexicano de Tecnologia del Agua (IMTA). 1995. Base de datos agrícola. FAO Mexico Internal Report. Project No. UTF/MEX/030/MEX.

Hernández, X.E., T. Levy and L. Arias. 1987. Hacia una evaluación de los recursos naturales renovables bajo el sistema de roza-tumba-quema en México. In. H.G. Lund, M. Caballero, R. Villarreal (eds). Evaluación de tierras y recursos para la planeación national de las zonas tropicales. USDA For. Serv. Gen. Tech. Rep. WO-39, Washington, DC, pp. 338-340.

Levy, T.S., E. Hernandez X. and L. Arias R. 1987. Aprovechamientos forestales en el sistema de R.T.Q. In. Memorias X Congreso de Btánica. Soc. Bot. Mex. Resúmen No. 27, 225 p.

Masera, O., M.J. Ordóñez and R. Dirzo. 1992. Carbon emissions and sequestration in forests: Case studies from seven developing countries. Volume 4. E.P.A. Climate Change Division. Lbl-32665.

Oyama Homma, A.K., R.T. Walker, F.N. Scatena, A.J Conto, R. Carvalho, A.C. Rocha, C.A. Palheta, A. Moreira. 1993. La dinámica de deforestación y quemadas en el amazonas: un análisis microeconómico. Rural Development Forestry Network. Documento 16c., 16 p.

Quintas, I. 1996. Extractor rápido de infomración climatológica (ERIC): Manual del Usuario. Instituto Mexicano de Tecnología del Agua. Colección proyectos IMTA.

Rzedowsky, J. 1978. Vegetación de México. Mexico city. LIMUSA.

Sánchez, P. 1976. Properties and management of soils in the tropics. John Wiley & Sons. New York.

Sánchez, V.A.S. and C.A. Ortiz S. 1991. El consumo de leña y su impacto sobre los suelos forestales del suroeste de Puebla. *Agrociencia Serie Rec. Nat. Renovables* 1(1):13-37.

Secretaria de Asentamiento Humanos y Obras Publicas (SAHOP). 1981. Plano de políticas ecológicas y plano de vegetación y uso del suelo escala 1:4,000,000. Mexico city.

Secretaria de Agricultura, Ganaderia y Desarrollo Rural (SAGAR). 1995. Anuario estadístico de la producción agrícola de los Estados Undos Mexicanos 1994. Secretaría de Agricultura Ganaderia y Desarrollo Rural. Subsecretaría de Planeación, Mexico city, 698 p.

Secretaria de Agricultura, Ganaderia y Desarrollo Rural (SAGAR). 1996. Compendio estadístico de la producción pecuaria de los Estados Unidos Mexicanos 1990-1994. Secretaría de Agricultura Ganaderia y Desarrollo Rural. Centro de Estadística Agropecuaria, Mexico city, 204 p.

Secretaria de Agricultura, Ganaderia y Desarrollo Rural (SAGAR). 1998. Programa de Praderas y agostaderos 1995-2000. Secretaría de Agricultura Ganaderia y

Desarrollo Rural. Comisión Técnica Consultu'iva de Coeficientes de Agostaderos, Mexico city, 100 p.

Secretaria de Agricultura y Recursos Hidraulicos (SARH). 1986. Información forestal de la República Mexicana. Subsecretaría de Desarrollo y Fomento Agropecuario y Forestal. INF. Mexico city. SARH.

Secretaria de Agricultura y Recursos Hidraulicos (SARH). 1990. Tabla de datos sobre perturbación forestal anual. Dirección de Protección Forestal. Mexico city. SARH.

Secretaria de Agricultura y Recursos Hidraulicos (SARH). 1992. Inventario nacional forestal de gran visión de México. Subsecretaría Forestal. Mexico city. SARH.

Secretaria de Agricultura y Recursos Hidraulicos (SARH). 1993. Sistema ejecutivo de datos básicos. Subsecretaría de Planeación. Mexico city. SARH.

Secretaria de Agricultura y Recursos Hidraulicos (SARH). 1994a. Inventario nacional forestal periódico. Subsecretaría Forestal y de Fauna Silvestre. Mexico city. SARH.

Secretaria de Agricultura y Recursos Hidraulicos (SARH). 1994b. Compendio estadístico de la producción pecuaria de los Estados Unidos Mexicanos 1989-1993. Secretaría de Agricultura y Recursos Hidráulicos. Subsecretaría de Planeación, Mexico city, 53 p.

Toledo, V.M. 1990. El proceso de ganaderización y la destrucción biológica y ecológica de México. *Medio Ambiente y Desarrollo en México* 1(2):191-222.

Vousden, N. 1973. Basic theoretical issues of resource depletion. *Journal of Economic Theory* 6(1):213-224.

Index